# Power Systems

For further volumes:
http://www.springer.com/4622

Thomas Worzyk

# Submarine Power Cables

## Design, Installation, Repair, Environmental Aspects

Springer

Thomas Worzyk
370 30 Rödeby/Sweden
worzyk.book@telia.com

Office:
ABB AB High Voltage Cables
371 23 Karlskrona/Sweden
thomas.worzyk@se.abb.com

ISSN 1612-1287 e-ISSN 1860-4676
ISBN 978-3-642-01269-3 e-ISBN 978-3-642-01270-9
DOI 10.1007/978-3-642-01270-9
Springer Dordrecht Heidelberg London New York

Library of Congress Control Number: 2009927023



*Cover design*: deblik, Berlin

Printed on acid-free paper

Springer is part of Springer Science+Business Media (www.springer.com)

# Preface

Submarine power cables have always been the unknown cousins of the submarine telecom cables. Telecom cables encircle the globe through all oceans in an enormous mesh, and have attracted public attention since the first Channel cable was laid in 1850. While the heroes of the first Atlantic telecom cable have been sung in numerous books, the feats of submarine power cable technology were rarely respected outside a small community.

However, submarine power cables have their own characteristics and peculiarities. Cable design has developed during the decades. The evolution of manufacturing and installation technology allows for projects that, 20 years ago, were considered impossible. New performance records in system length, water depth, transmission power, or profits from the operation of submarine power cables are set every few years. Owing to this development, more submarine power cable projects are economically viable, or even possible. New applications for submarine power cables appear while known applications are pursued on a larger scale.

Submarine power cables draw the attention of larger circles. Ten years ago only power utility engineers and cable ship crews were interested in submarine power cables. Possibly some fishermen uttered unprintable comments on them when they had entangled their fishing gear. Today, in the wake of new applications, submarine power cables have come into the focus of larger groups. The engineers of oil-and-gas companies have been interested a while, and increasing offshore activities, as well as the need for more power supply from shore to platform have promoted their involvement in the field. Most of all the fast development of offshore wind power brought submarine power cables to the attention of new quarters outside the power utilities. Investors and project developers need knowledge on the subject in order to analyse the technical and economical feasibility of offshore wind farms, as the cable costs can make up to forty per cent of the project costs, and cable installation can have a substantial impact on the overall schedule. Great commercial opportunities in power trading triggered the curiosity of Transmission System Operators (TSO) and power traders about opportunities and limitations of submarine power transmission lines. Regulators and authorities are getting more involved into submarine power cables, as permit applications continue to roll in. Also environmental groups and organisations now have to deal with submarine power cables, which sometimes are installed in sensitive areas. Satisfying enough, more students of all levels are getting involved into submarine power transmission as they study the possibilities of renewables. Even lawyers have developed an occupational interest in these cables.

The idea to prepare this book actually came when I received many questions from the new audience, which demonstrated that there was need for a comprehensive introduction into the matter. For those who want to dig deeper into the issue it is not always easy to find relevant information. Much needed information about design and installation is scarce. Most textbooks on cables deal with land cables and leave only little space to submarine power cables, sometimes relegating them into their "Miscellaneous" chapter. Other cable textbooks are simply outdated neglecting state-of-the-art manufacturing and installation techniques.

When I put this book together, I had to face several difficulties. Evaluating published reports and peer discussions, it became clear that the willingness for sharing experiences has decreased strongly. Information on submarine power cable operations have been shared in journal papers quite openly until the 1990s, while reports about recent projects rarely address "lessons learned". At a time when investor relations, quarterly reports, and insurance issues prevail, few cable operators like to go public with accounts of detailed technical issues, problems, and fixes. However, the entire community of investors, grid operators, manufacturers, installers etc. would profit from an openhearted dialogue.

This book aims at conveying a basic knowledge on the design, manufacture, and installation of submarine power cables. In 1903, cable engineer Dr. C. Baur wrote in the preface to his cable book: "the amount of knowledge related to the electric cable is so large that no single man can comprehend it". This must be even more correct for submarine power cables. I must most humbly admit that I could only grasp and compile a fraction of the knowledge on the subject, relying on the assistance and help of many colleagues. For those who want to go further the reference lists may be helpful.

I owe a lot of gratitude to those who assisted in the development of this book. Most of all, at this prominent place of the book, I want to take the opportunity to thank my dear wife for her enormous patience and my sons for their valuable input from their point of view.

# Acknowledgments

I am indebted to many people and companies who helped me to bring this book on its way and into a reasonable form. I want to make a point of thanking all of them at this occasion.

Although the production of this book was my private adventure, my employer ABB/Sweden provided me with a grant for literature and xerox copies. My colleagues let me participate in their enormous and priceless knowledge and experience, just to contribute to this book. Peter Sunnegardh, Anders Gustafsson and Johan Karlstrand supported me with their most valuable comments.

Other cable manufactures gave me indepth information. I want to express my gratitude to Prysmian/Italy, Norddeutsche Seekabelwerke NSW/Germany and NKT Cables/Denmark for their generous assistance.

The power utilities Statnett/Norway, Long Island Power Authority/USA, and Baltic Cable/Sweden contributed with information on cable service experiences and other. HT Suen, John Savio and Jan Brewitz were ever so cooperative.

A book cannot be completed without the involvement of certain individuals who are conducive with illustrations, knowledge, anecdotes, good advice, or something else – or a bit of all of this. I would like to thank Lars Aksel Solberg and Janne Stark for a good story, Bill Burns for background info, Walter Paul and Jaeyoung (Jay) Lee for most interesting facts.

Lydia Stark and Stephen Wigginton helped me to improve the writing language.

The challenging task of transforming my writing into readable English was performed by Jochen U. Frankl, of Keryx Translations/Germany.

A number of other companies helped me out with much needed photographs. Commercial and non-commercial marine organisations, from all around the world, have played a decisive role in the improvement of this book. The list is here:

ABB Oy, Finland; ABB AB, Sweden; Baltic Cable AB, Sweden; Borealis, Sweden; IHC Engineering Business Ltd, UK; International Cable Protection Committee, UK; LD Travocean, France; Long Island Power Authority, USA; Minerals Management Service, U.S. Department of the Interior, USA; National Institute of Water & Atmospheric Research Ltd, New Zealand; NKT Cables A/S, Denmark; NSW, Germany; Oceanteam Power and Umbilicals, UK; Prysmian, Italy; Statnett, Norway; Submarine Telecom Forum, UK; Trelleborg, UK; Woods Hole Oceanographic Institution, USA.

# Contents

# About the Author

Thomas Worzyk, born in Germany in 1957, received his M.Sc. degree in Physics from the Leibniz Universität Hannover, Germany in 1983. He joined one of the largest electric power product companies in Sweden, in 1984. After some years in the circuit-breaker R&D department, he moved on into the field of high-voltage cables, when a new cable factory was set up in the town of Karlskrona, Sweden.

Worzyk has acquired a deep knowledge in research and design aspects of power cables, especially submarine. He has been the Chief Engineer in several large cable projects, both underground and submarine. He spent almost a decade with the development and implementation of the longest power cable link of all classes, which was inaugurated in 2008.

During the work with this book, Worzyk has discovered his inclination to the history of submarine power cables. "From history, we sometimes can learn things cheaper and faster than from research projects," he concludes.

Worzyk has published a number of journal articles and is holder of patents in his field.

# Abbreviations

| | |
|---|---|
| 3C | Three-Core Cable |
| 1C | Single-Core Cable |
| AHT | Anchor Handling Tug |
| AIS | Automatic Identification System |
| CDVC | Cable Depending Voltage Control |
| Cigré | Conseil International Des Grands Réseaux Électriques |
| CLPS | Cable Load Prediction System |
| DTS | a) Distributed Temperature Measurement System<br>b) Desktop Study |
| EIA | Environmental Impact Assessment |
| GIL | Gas Insulated (Transmission) Line |
| GIS | Gas Insulated Switchgear |
| IEC | International Electrical Committee |
| LCA | Life Cycle Assessment |
| LIWL | Lightning Impulse Withstand Level |
| LPOF | Low Pressure Oil Filled |
| MBR | Minimum Bending Radius |
| MI | Mass-Impregnated |
| OTDR | Optical Time Domain Reflectometry |
| OWP | Offshore Wind Park |
| PD | Partial Discharge |
| PLGR | Pre-Lay Grapnel Run |
| ROV | Remote Operated Vehicle |
| RTTR | Real Time Thermal Rating |
| SC | Single-Core Cable |
| SCADA | Supervisory Control And Data Aquisition |
| SCFF | Self-Contained Fluid Filled |
| SCOF | Self-Contained Oil Filled |
| SIWL | Switching Impulse Withstand Level |
| STRI | Swedish Transmission Research Institute |
| TD | Touch Down |
| TDR | Time Domain Reflectometry |
| TSO | Transmission System Operator |
| VIV | Vortex Induced Vibration |
| VMS | Vessel Monitoring System |
| WTG | Wind Turbine Generator |
| XLPE | Cross-Linked Polyethylene |

# Chapter 1
# Applications of Submarine Power Cables

## Contents



Submarine power cables have been around for more than a century and the major use has shifted through the decades. In the early times, submarine power cables were used to supply isolated offshore facilities such as lighthouses, infirmary ships, etc. Later, the power supply of near-shore islands was the main objective of submarine power cables. The connection of autonomous power grids for the sake of better stability and resource utilisation has been pursued since the sixties of the last century. In modern days, the connection of offshore facilities is again in the focus. Oil and gas production units ask for shore-generated power, while offshore wind parks (OWP) need to bring their precious green power to the onshore grids.

## 1.1 Power Supply to Islands

Islands located closely to the mainland can be connected to the mainland grid by submarine power cables. This is normally done with medium-voltage a.c. cables ($\leq$52 kV) and a transmission power of 10–30 MW per cable. The submarine power cables replace island-stationed, often-inefficient power generation such as diesel generators. The maximum economic length of these cables is 10–30 km. In response to increased power demand of the island, additional cables are often laid in different routes to reduce risks and increase the power availability on the island. The island supply can be secured by other cables even in the case of a cable failure. The archipelago of the North Frisian Islands in Northern Germany was connected to

T. Worzyk, *Submarine Power Cables*, Power Systems,
DOI 10.1007/978-3-642-01270-9_1, © Springer-Verlag Berlin Heidelberg 2009

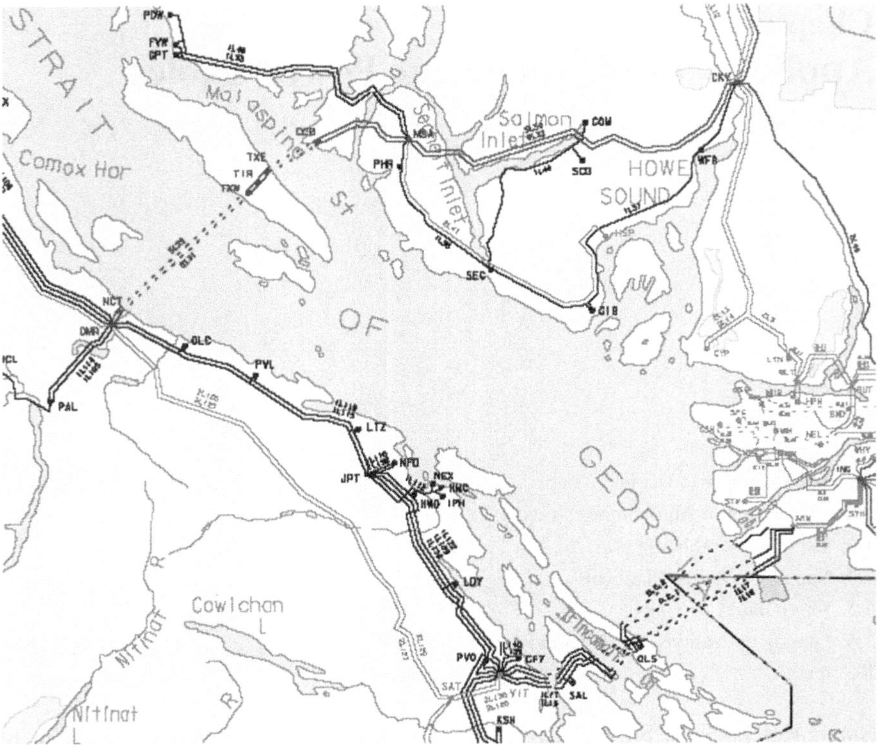


**Fig. 1.1** Submarine power cables supplying Vancouver Island, Canada. The island (*down left*) was first connected to the mainland (*upper right*) with a 138 kV system (*down right*). Increasing power demand necessitated more cable systems, both a.c. and d.c. [2]

the mainland by a grid of 20 kV submarine cables starting in 1944 [1]. Other islands like Vancouver Island, Canada, and Long Island, NY, USA, have a large population and massive power demands requiring a number of extra high voltage submarine power cables circuits (Fig. 1.1).

The supply of distant islands is difficult as losses in a.c. cables increase dramatically with distance. Islands located more than 50 km from the mainland have not been connected for a long time. The German tourist island of Heligoland will be connected to the mainland only later this year (2009). The Swedish island of Gotland is some 100 km off the Swedish mainland and had to rely on inefficient diesel power generation for a long time. The island wasn't connected to the mainland grid until 1954, when the first submarine HVDC cable was installed. The canal islands of Jersey and Guernsey were connected to the French mainland in 1987 and again 2000, and the Italian islands of Sicily and Sardinia were connected to the Italian mainland by submarine power cables.

Islands with an autonomous power supply are sometimes connected to the mainland or neighbour islands in order to increase the power availability. The submarine cable has the function of a spare power plant providing emergency power in case of a local generator outage.

Owing to the geographical character, some countries, such as Norway, the Philippines, Japan, and other countries, have a long tradition of installing submarine power cables between their numerous islands.

## 1.2 Connection of Autonomous Grids

Since the advent of powerful submarine cables many grids have been interconnected, using different techniques. Submarine cables connect grids of different countries (To name a few: UK – France, Sweden – Germany, Denmark – Sweden, Morocco – Spain, Greece – Italy). Using HVDC techniques, regions with different frequency control can be linked (Store Belt in Denmark, Sweden – Germany, Norway – the Netherlands). In the mentioned cases, both sides have a 50 Hz grid but the frequency is controlled in different ways, rendering asynchronous grids. The interconnection of national grids is a prime objective of the European Community.

The connection of autonomous grids can have various purposes and objectives:

- The load peaks of the connected countries/grids occur at different times of the day due to different time zones or due to different habits of electricity use. With the submarine cable transmission it is possible to share generation capacities to meet the power demand.
- The connected countries/grids may have different electricity generation mixes with differences in availability and pricing patterns. Many HVDC cables connect the NordEl grid system with its abundant hydropower to the European continental power grid (UCTE), which mostly relies on fossil and nuclear power. "Green power" can be traded across borders.
- Each power grid needs a certain amount of "spinning reserve", that is generation capacity that can be switched to the grid within minutes. Traditionally, generators (hydro or thermal) provide spinning reserve while running idly. It is obvious that a submarine power cable making a connection to another grid is a more efficient and maintenance-free spinning reserve, which can be used to dispatch power flow in either direction within minutes.
- Since the deregulation of the power markets, the pricing pattern is highly volatile. Power traders can use submarine power cables to earn money by exploiting the price difference in the connected countries/grids. Sometimes they change the direction of power flow several times a day responding to the price fluctuations [3].

Submarine HVDC power cables of extreme length have been used to connect distant autonomous grids: The longest existing HVDC submarine cable system is a pair of 580 km cables (NorNed), and a number of 200+ km systems are in successful operation (Baltic Cable, Swepol, Bass Link) or under construction (BritNed, SAPEI).

## 1.3 Offshore Wind Farms

Offshore wind farms (OWP) consists of a number of wind turbine generators (WTG). The distance between the WTGs is 300–800 m. A grid of submarine cables interconnects the WTG and bring the power to shore. The in-field cables are three-phase medium-voltage cables (10–36 kV) with polymeric insulation. The connection to shore can be achieved with medium-voltage cables for distances up to about 10 km.

For OWP with many turbines, large output power, or large distance to the shore, the losses in a medium voltage transmission to land would be quite high and a high-voltage connection to land would be more economic. In larger offshore parks the WTG are connected to an offshore platform that carries a step-up transformer. From the transformer platform, a submarine power cable (export cable ) sends the power to shore. Three-phase cables with >100 kV operating voltage serve most often as export cables for distances exceeding 30 km (Fig. 1.2).

Export cables may also operate with HVDC, which requires converter stations both offshore and onshore. The erection of an offshore converter station on a platform is expensive and can be motivated only when large amounts of power must be transmitted over a long distance.

The use of a.c. frequencies lower than 50 Hz for WTG has been considered, in order to reduce capacitive losses in the cables, and mechanical requirements for the WTGs.

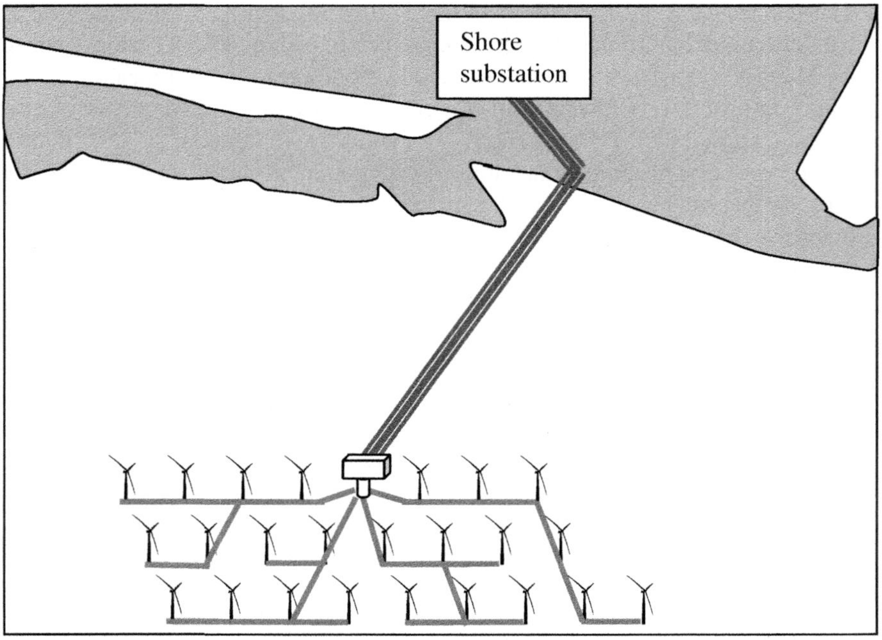


**Fig. 1.2** Layout of an offshore wind farm with 34.5 kV in-field cables and 138 kV export cable between the OWP and the shore line

## 1.4 Supply of Marine Platforms

Production platforms in the offshore oil and gas business have a large power demand to extract hydrocarbons from wells. Energy use covers a range of activities including driving pumps to extract hydrocarbons and to re-inject water for enhanced oil recovery, heating the output stream to allow separation of the oil, gas and water, powering compressors and pumps for transporting oil and gas through pipelines to processing plants or to shore, and supply of the electricity needed for on-site operations and living quarters.

Energy needs vary widely according to local circumstances and operational conditions. The electric power for many platforms is generated from locally produced gas in steam plants or gas turbines at rather low efficiency. The power plants require precious space while their operation and maintenance staff needs additional accommodation and transport. All this makes the onboard power production expensive. As the power need increases, it becomes viable to connect the platform to onshore grids by submarine power cables.

In Norway, the offshore industry produced a quarter of the country's total $CO_2$ emission in 2006. As Norway has vast resources of hydropower, there is a large potential to reduce the country's $CO_2$ emission by connecting platforms to the onshore grid (Fig. 1.3).

**Fig. 1.3** The Norwegian gas platform "Troll A" receives its electric power supply from a pair of HVDC cables (Courtesy of ABB, Sweden)

For these reasons, an increasing number of platform operators invest in submarine power cables for the power supply of offshore platforms. The supply of floating platforms is a special challenge as the repeated movements, induced by wind and waves, require certain design considerations.

## 1.5 Short-Haul Crossings

Hundreds of submarine power cables have been installed to transport power across rivers, channels, straits, fjords, or bays. Though overhead lines can be used for crossings up to 3 km (e.g. Messina Strait, Italy) in many cases submarine cables have been chosen instead of overhead lines. The invisibility of cables is important in tourist resorts and natural resources. Allegedly, for the crossing of a 500 kV d.c. power transmission over the St. Lawrence River, Canada, a 5.1 km submarine cable system was used to avoid optical impact. Cables do not restrict the height of ships passing the river or strait. The overhead lines over the Ems River in Germany have to be shut down each time when an upstream shipyard transports its new mega-size cruising ships to the North Sea. When the lines were switched off the grid in 2006 for the passage of the "M/S Norwegian Star" a galloping instability blacked out millions of households in Europe. Sub-river cables have been discussed for this river crossing but were not yet realized.

Also the lifetime costs of a maintenance-free cable can be lower than that of an overhead line threatened by storms, salt and ice deposition, etc.

In short-haul applications, submarine cables with very high voltage can be used when no joints are required. Often cables are available at higher voltages than the corresponding joints.

Crossings of some 1400–1800 m can be realized with trenchless horizontal directional drilling methods from the shoreline. The bore is lined with pipes and the cables would simply be pulled through the pipes. Longer crossings would require ordinary submarine cable laying techniques.

## 1.6 Other Applications of Submarine Power Cables

There are many niche applications of submarine power cables that cannot be covered in this book. The submarine cables used here are in the medium-voltage or low-voltage range. Here are mentioned a few:

*Oil and gas production cables.* As production wells are established in ever larger depths of water, more sophisticated equipment such as submersible pumps and compressors of all kind is placed on the seafloor. Figure 1.4 illustrates different oil and gas installations all requiring submarine power cables.

*Umbilicals* are armoured flexible assemblies containing anything from power cable cores, signal cables, fluid conduits, hydraulic lines etc in the same

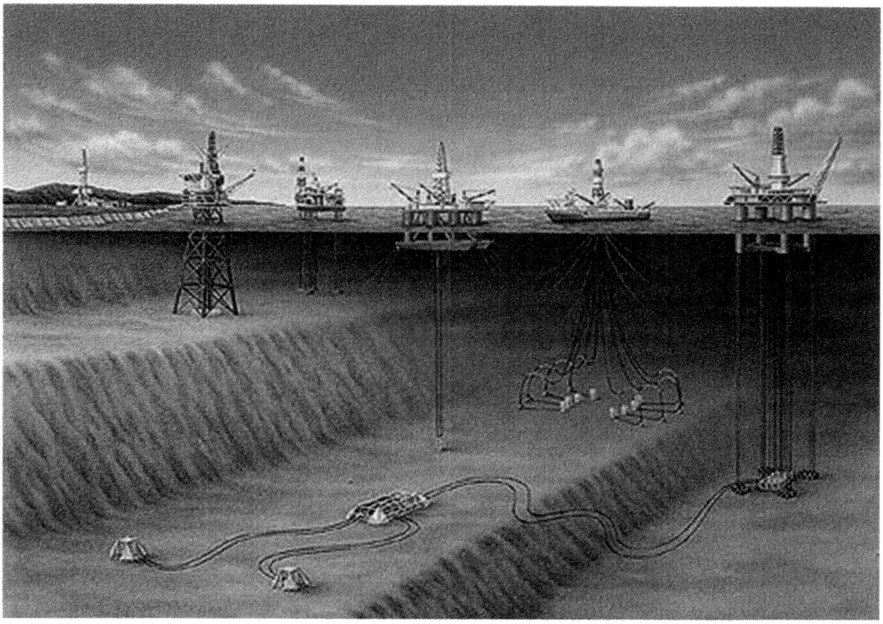

**Fig. 1.4** Various oil and gas installations requiring submarine power cables. (Courtesy of Minerals Management Service, US Department of the Interior, www.mms.gov)

cable. They serve oil and gas equipment on the seafloor and remotely operated subsea vehicles (ROV).

*Pipeline heating cables.* Submarine pipelines sometimes need electric heating to prevent the formation of wax and hydrate deposits. The pipe itself is used as the heating element, and the power supply is accomplished by XLPE-insulated power cables with large conductor size and without metallic water barrier [4].

*Subsea observatories.* Tsunami pre-warning systems and military reconnaissance arrays require underwater power. The increasing focus on oceanographic research has triggered the installation of subsea automatic research stations to collect data. All these must have reliable submarine power supply [5].

## References

1. Doyen H et al. (1989). Experiences with Different Cable Designs and Laying Methods in Conjunction with the Power Supply of the Islands in the North and Baltic Sea, CIRED 1989.
2. Suen H (2006). Vancouver Island – Transmission Reinforcement Project 242 kV Submarine Cable. Presentation at the IEEE-ICC Meeting, St. Petersburg, Fla, USA, 1 November 2006.
3. Hammonds T J et al. (2000). Role of HVDC Transmission in Future Energy Development, IEEE Power Engineering Review, February 2000, p. 10–25.

4. Hvidsten S et al. (2004). HV Cable Design Applicable for Direct Electrical Heating of Very Long Flowlines. Proceedings of OCEANS '04 and MTTS/IEEE TECHNO-OCEAN '04, Kobe, Japan.
5. Shepherd K et al. (2007). Observatory Cable Laying Systems. Oceans 2007, September 29–October 4, 2007, p. 1–5. ISBN 978-0933957-35-0.

# Chapter 2
# Submarine Power Cables and Their Design Elements

## Contents




T. Worzyk, *Submarine Power Cables*, Power Systems,
DOI 10.1007/978-3-642-01270-9_2, © Springer-Verlag Berlin Heidelberg 2009

Many different shapes and styles of submarine power cables have been invented, developed, manufactured, tested, and installed during more than hundred years. The showrooms of the cable manufacturers display a fascinating variety of designs, all made with contemporary engineering art and entrepreneurship. There are success stories, and there are sad stories. This book concentrates on cable types that are produced and installed today (2009). Other cable types, which still might be in operation and maybe subject to repair operations are mentioned and described briefly. This chapter tries to describe different submarine cable types and their construction elements without going too deep into formulae.

Five very typical submarine cable designs and their application are portrayed in Sect. 2.10 at the end of this chapter.

## 2.1 The Conductor

The current-carrying conductors of submarine power cables are made of copper or aluminium. Even though copper is more expensive than aluminium in relation to the current-carrying capability, the majority of submarine power cables have copper conductors. Copper allows a smaller cross section and hence requires less material for the outer layers such as lead and steel wires. However, there are occasions where aluminium is the better solution. There is no given best choice as the costs vary strongly with a volatile metal market. The limited corrosion resistance of aluminium is sometimes quoted in favour of copper for submarine power cables. However, when seawater can cause corrosion in the conductor, it has already penetrated the insulation, and the cable must be repaired or replaced, regardless of the conductor material.

For some cable projects it makes sense to use copper for some parts of the route, and aluminium for others. The HVDC cables of the Estlink project have copper conductors in the submarine part and aluminium in the land section [1]. It is also possible to use submarine cables with aluminium conductor for the deeper cold part of the submarine cable route and "copper ends" for the warmer parts buried close to the beach [2]. Copper and aluminium conductors can be jointed together.

Conductors of submarine power cable are made in various shapes (cf. Fig. 2.1). The most common shapes are explained below.

### *2.1.1 Solid Conductor*

The conductor consists of a single massive wire. This design is used for cross sections up to about 400 mm$^2$. The manufacturing is easy and the conductor has

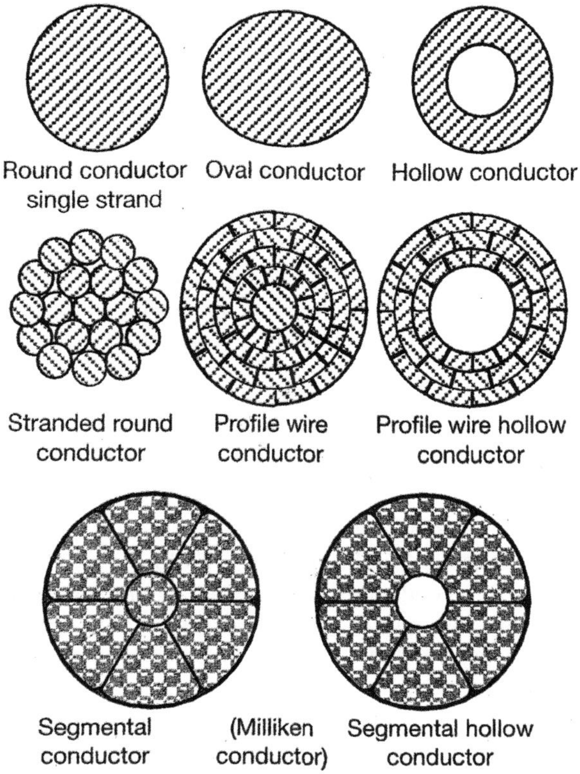


**Fig. 2.1** Conductor design

naturally good longitudinal water tightness, a property most often required for submarine cables. Sometimes an extruded insulation tends to slip on the smooth conductor surface thus creating a shrink-back tendency of the insulation when cutting the cable for termination or jointing. This phenomenon depends largely on the extrusion technique for the insulation. Knurling or similar surface treating methods on the conductor can enhance friction and reduce shrink-back problems.

As the cross-section of solid conductors is limited to about 400 mm$^2$, they are not used for cables >150 kV, which usually have larger cross sections. For low voltage cables with three or four cores in one cable, also segment-shaped solid conductors are used.

## *2.1.2 Conductors Stranded from Round Wires*

Most conductors for submarine power cables are stranded from round wires. The wires are laid up in the stranding machines in layers. The conductor is compressed by the action of dies or roller sets, either after each layer or at the end of the stranding machine. The compressing reduces the interstices between the strands. Conductors from compressed round wires can achieve a filling factor of 92%. Since the wires are cold worked by the compression, the electric conductivity of the material is reduced.

There are two different stranding directions: righthand or left hand. Sometimes they are denominated as Z-lay or S-lay, depending on the visual impression of the spiralled strands. Imagine that the letter is projected onto the lay and find out if the central line of the letter goes in parallel with the strand lay. The sketch shows a Z-laid conductor. Most often, the layers of stranded conductors have alternatively S- and Z-lay, but it is possible to use any mixture of lay directions for better stability of the conductor or other production reasons. For any practical reason, the conductivity of the conductor is not depending on the sequence of S- and Z-lay. If the cable is to be coiled in clockwise direction the outer conductor layer should be laid right-hand (Z-lay) so that the conductor contracts during coiling [3]. A conductor expanding under the influence of the coiling stress would possibly infringe with the insulation. In paper-insulated cables the innermost paper layers can burst.

Conductors with compressed round wires are equally useful for a.c. and d.c. applications. Most often, they are designed according to the industrial standard IEC 60228, Class 2. For large cross sections or transmission power, d.c. cables would be more economic with profiled wire conductors, and a.c. cables would be more economic with Milliken conductors which are described later.

In large conductors for a.c. cable additional magnetic losses can be generated reducing the ampacity[1] of the cable. The "proximity effect" in a conductor is caused by the magnetic field of neighbour cables and tends to "push" the current flow lines into a certain part of the conductor. Individual wires or groups of wires can be insulated electrically to reduce the "proximity effect". The insulation can be accomplished by paper or plastic strips, or by varnish coatings of the wires. Another magnetic effect, the "skin effect" cannot be reduced by this method in normal stranded conductors.

Tinned copper wires have been used frequently for conductor manufacturing in earlier times, but this concept has not been pursued in the recent years.

### 2.1.3 *Profiled Wire Conductors*

The conductors are composed from cake-piece-shaped wire cross sections, sometimes also called keystone wire conductors. In the stranding machine, the wires (strands) combine perfectly to a circular conductor. Filling grades up to 96% can be achieved, probably even more. The conductor surface is very smooth, which is extremely useful during the further production process. Large HVDC submarine cables are most often produced with profiled wire conductors. The standard IEC 60228 does not cover conductors with profiled wires.

---

[1] The word ampacity denotes the current carrying capability of a cable and has been coined by W. A. DelMar of Phelps Dodge Wire&Cables in 1951 [31]

Profiled copper wires can be produced in almost any shape using the Conform method, which is a type of extruding. The shaping is done without cold working and the resulting wire has the excellent conductivity of annealed copper.

### 2.1.4 Hollow Conductors for Oil-Filled Cables

Oil-filled or fluid-filled cables (also called LPOF, low-pressure oil-filled, or SCFF, self-contained fluid-filled) are filled with low-viscosity oil. They contain a central duct in the conductor to allow for oil flow in connection with thermal expansion and pressure supply from the cable terminations.

Some designs have a central metallic support helix, which prevents the conductor wires from collapsing into the central duct. The helix is visible in Fig. 2.2.

Hollow conductors can also be made from profiled wires. Slot-and-key shaped profiled wires engage when stranded to form a self-stabilizing vault (cf. Fig. 2.3). In this case, the central helix can be omitted. A grooved surface between the wires helps to allow for sufficient oil flow between insulation and central duct.

### 2.1.5 Milliken Conductor

The alternating magnetic field from the a.c. current in a conductor produces an electromagnetic force (emf) that concentrate the current flow lines into the peripheral parts of the conductor (the "skin"). The current squeezes into the outer part

**Fig. 2.2** LPOF cable made by Taihan

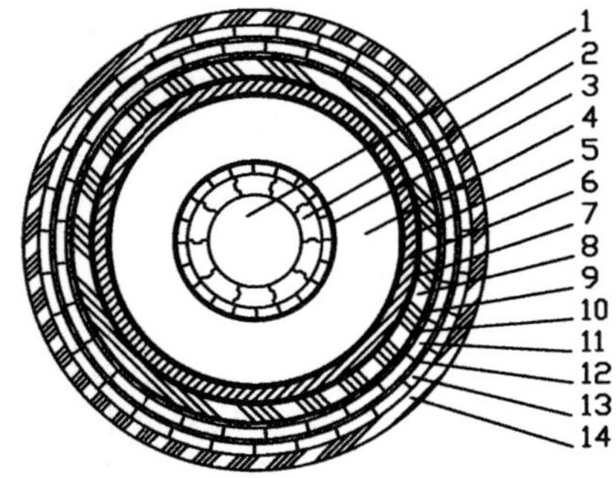


| Item | Component | Item | Component |
|---|---|---|---|
| 1 | Oil Duct | 8 | Anti-corrosion PE jacket |
| 2 | Conductor, self supporting segmental strips of copper | 9 | Antiteredo protection, Cu tapes |
| 3 | Conductor screen | 10 | Bedding |
| 4 | Insulation, paper tapes impregnated with low viscosity fluid | 11 | Copper flat wires wire armor |
| 5 | Core screen | 12 | Bedding |
| 6 | Lead sheath | 13 | Copper flat wires wire armor |
| 7 | Reinforcement, bronze tapes | 14 | Serving |

**Fig. 2.3** Hollow conductor with slot-and-key shaped wires (Courtesy of Prysmian, Italy)

of the conductor. The current density in the inner conductor part decreases and the inner part of the conductor is less useful for the current transport. The skin effect reduces the ampacity of large conductors for alternating current substantially. H. Milliken filed two patents in the 1930s describing a conductor design with reduced skin effect [4, 5]. For the manufacturing of a Milliken conductor, the factory starts with standard stranded sub-conductors from round wires. These sub-conductors are now rolled into a triangular cake-slice shape, and then are pre-twisted. Finally, a number of these sub-conductors are laid up into a round conductor. The result is obvious from Fig. 2.2. A Milliken conductor may or may not have a central duct or a central wire. Milliken conductors are also known as segmental conductors.

In a standard conductor, the distance to the conductor axis of each individual wire is constant along the conductor. The individual wire accumulates the emf induced at its particular distance to the conductor axis. In contrary, in the Milliken conductor each individual wire changes its radial position from close to the centre to far from the centre as the wire proceeds along the conductor. The emf at the centre and the emf at the periphery have opposite signs and cancel each other partly. This reduces the skin effect considerably. The larger the number of segment strands in the Milliken conductor, the better the reduction of the skin effect. In practice, segment numbers of 5 up to 6 are common. However, cables with up to nine segments have been designed [6].

For the full exploitation of the intricate current flow pattern in a Milliken conductor, it is important that the current cannot "jump" between the wires rather than following the wire. The insulating oil film between the individual wires helps reducing the skin effect. The application of extra insulation between the individual wires can reduce the skin effect even further [7]. This insulation can be applied between the individual segments, between the wire layers within each segment, between individual wires or any combination thereof. It is also known to insulate only a fraction of all individual wires. The insulation can be made with thin paper strips or lacquer on the wires.

The higher production costs of Milliken conductors motivate their use only for large cross sections ($\geq$1200 $mm^2$ copper).

### 2.1.6 Conductor Resistance

The electric resistance is the most important property of the cable conductor. The specific resistivity of the conductor material is temperature dependent:

$$R_\theta = R_{20} \cdot (1 + \alpha(\theta - 20))$$

with these symbols:

| | |
|---|---|
| $R_\theta$ | specific resistivity at temperature $\theta$ (°C) |
| $R_{20}$ | specific resistivity at 20°C |
| $\alpha$ | Temperature coefficient of the electric resistivity |

The specific resistance and the thermal coefficient for copper and aluminium are given in Table 2.1. Both the specific resistivity and the temperature coefficient are depending on the purity of the metal. The effective resistance of a given conductor is not only depending on the conductor cross-section but also on the lay length of the strands because the current has to travel a longer path in a laid-up conductor. The number of strands and the compression during lay-up also influences the resistance. The presence of water-blocking compounds in the conductor can increase the resistance as contact points between individual wires vanish. During the manufacturing, the wires are subjected to different degrees of cold working. Cold worked copper and aluminium have a higher specific resistivity than annealed metal. Power cable specifications usually not only specify the conductor cross-section but also the

**Table 2.1** Specific resistivity and its temperature coefficient

| | Copper | Aluminium |
|---|---|---|
| $R_{20}$: Specific electric resistivity @20°C, $\Omega mm^2/m$ | 0,01786 | 0,02874 |
| $\alpha$: Thermal coefficient of the specific electric resistivity @20°C, 1/K | 0,00392 | 0,0042 |

conductor resistance. The standard IEC 60228 lists conductor resistance values that include the influence of the strand lay length (cf. Table 3.1 in Chap. 3).

With the Conform® Extrusion process, shaped wires can be made without cold working resulting in better conductivity values than those for compressed round wires. The resulting profile wire conductor may have a smaller conductor than a compressed wire conductor to achieve a specified resistance.

The purchaser of a power cable is not primarily interested in a particular cross section, but in obtaining a cable with a certain resistivity (or conductivity). Ordering a cable with 1600 $mm^2$ copper, the purchaser actually wants to have a cable with a resistance not more than 0.0113 Ω/km according to the IEC60228. This value is the basis for the system owner's loss and ampacity calculations. However, it is up to the skills of the manufacturer to achieve the resistance requirements with as little copper as possible, using the best art of choosing wire production method, lay length, compression factors, raw material grade, etc.

Standardized conductor sizes have been tabulated both in the IEC60228 standard and many national standards. However, submarine cables are often tailor-made products for a specific project. It is quite common to determine a suitable cross section area from the actual needs rather than from the standard values. Odd values such as 790, and 1410 $mm^2$ have been chosen to meet real needs. Only in very rare cases, submarine power cables can be purchased from stock at standardized cross section values.

### *2.1.7 Watertightness of Conductors*

Longitudinal water tightness is often required for submarine power cables to prevent water from migration into the cable after a cable fault. Also, water ingression from defect end caps during transport or installation must be prevented. For this purpose, swelling agents in the shape of powder, tapes or yarns are inserted in between the conductor layers. Upon contact with water they swell considerably and block the passage for water efficiently. Most swelling agents work much better in contact with fresh water rather than salt water. Other hydrophobic compounds are used, too, to stop water migration. Petrojelly is a vaseline-based material serving the same purpose. It has been discussed to extrude a polymeric matrix accommodating the conductor wires and providing high longitudinal water tightness. Oil-filled and mass-impregnated cables are considered having longitudinal water tightness without additional measures to be employed.

### *2.1.8 Superconducting Conductors*

No other conductor materials than copper or aluminium are used for commercial submarine cable manufacturing today. Research institutes investigate the possibility to make conductors with very high conductivity from carbon nano-tubes.

After the discovery of high-temperature superconductivity in 1986, a number of oxide compositions have been found with a critical temperature for superconductivity over 77 K, which is the lowest temperature for liquid nitrogene. Superconductivity came into reach for many practical applications. Material systems like Y–Ba–Cu–O, Bi–Sr–Ca–Cu–O und Tl–Ba–Ca–Cu–O are in the frontline. Very soon, laboratory samples of cryogenic superconducting cables were devised. For cable applications, the superconducting materials are shaped into thin ribbons to be applied onto a copper substrate. At the end, the ribbons can be crafted into cable conductors with enormous ampacity when under the critical temperature. Experimental cables have been produced and a large effort of dielectric research on cryogenic materials is done. Still, there are some major drawbacks: The power needed for the cooling process is only little less than the saved losses. For submarine applications there is still no suitable cryogenic cable available.

## 2.2 The Insulation System

The cable insulation provides an effective barrier between potential surfaces with an extreme potential difference. It is of utmost importance that the insulation system is absolutely clean and even. Furthermore, the insulation wall must be mechanically robust, and resistant to temperature and aging. During more than 150 years of submarine electric cables (telecom and power) many different insulation materials have been tried, developed, rejected or approved. Insulation materials for submarine cables are in no way different to insulation materials for land cables; however, conditions of production and application might be different. Today, submarine power cables for medium and high voltage are manufactured with only a few different insulation materials. In-depth information on the dielectric properties and chemical structure of cable insulation materials can be found elsewhere [8]. Here the most important characteristics of some insulation materials are described. Some properties of relevant insulation materials are compiled in tables in Chap. 3.

### *2.2.1 Polyethylene*

Polyethylene is a hydrocarbon consisting of long chains of $CH_3$–$(CH_2)_n$–$CH_3$ molecules. It contains nothing but carbon and hydrogen. The chains may be branched or not. The material is nonpolar and semicrystalline. It is thermoplastic and can theoretically be re-melted. For electric insulation it is available in different density ranges such as LDPE (low density), MDPE (medium-density), and HDPE (high-density). The density of all these varieties is between 0.9 and 0.97 g/cm$^3$. PE has lower dissipation factor and lower dielectric losses than paper insulation and was used for 63 kV cables since the early 1960s. In France, LDPE was used successfully for up to 500 kV cables in the 1990s. However, the author does not know if thermoplastic PE cables have been manufactured for submarine application. Because of the

**Table 2.2** Operational temperature of cable insulation materials

| | Operating temperature | Short-circuit temperature |
|---|---|---|
| LDPE | 70°C | 125°C |
| XLPE | 90°C | 250°C |
| EPR | 90°C | 250°C |
| Mass-paper | 50–55°C | |
| Oil-paper | 85–90°C | |

limited conductor temperature in PE cables (70–80°C), PE has subsequently been replaced by cross-linked PE (XLPE) which stands 90°C and short-circuit temperatures well above 200°C. However, as cable operators today try to avoid large losses associated with a hot conductor the thermoplastic PE insulation might experience a revival (Table 2.2).

### *2.2.2 Cross-Linked Polyethylene*

Cross-linked polyethylene (XLPE) has been used for submarine cables since 1973, and even earlier for land cables. Starting with polyethylene, XLPE is made by cross-linking the long molecular chains of LDPE to form a three-dimension network. The cross-linking is irreversible and prevents the polymer from melting at elevated temperatures. Thermoplastic polyethylene (PE) is getting soft and eventually melts when heated to 80–110°C depending on the density. In the contrary, XLPE is stable at far higher temperatures. Instead of by melting, it is destroyed by pyrolysis above 300°C.

Organic peroxides are being used as process initiators for the cross-linking process. They are added to the raw material already at the compounder's factory. The raw material is extruded around the conductor in the extruder head. The cross-linking takes place in a tube under inert atmosphere with high pressure and high temperature immediately after the extrusion head.

The gaseous by-products tend to form bubbles inside the still soft XLPE in the cross-linking zone of the extruder. To suppress the generation of gas bubbles, the extruded cable must be transported further in a pressurized tube until the cooling XLPE has gained sufficient strength. The cooling must be slow to relieve internal stress. The total travel length of the XLPE cable in the cross-lining zone, the cooling zone and the relaxation zone can be more than 150 m.

In vertical extrusion lines and in catenary lines, the contact between the tube wall and the XLPE insulation can be avoided, which leads to a very high surface quality. A vertical line prevents also the forming of excentrical insulation under the influence of gravity. In less common processes the cross-linking can be achieved by silane agents or electron ray irradiation.

XLPE is a first-choice insulation material for submarine cables. In the early days of XLPE cable manufacturing, the material got a bad reputation due to its perceived sensitivity for water. Under the combined influence of water, electric field and impurities, water-treeing could be initiated, a phenomenon involving tree-like damage structures inside the insulation. The "trees" would grow and finally cause an electric breakdown in some cases. Water-treeing has been investigated widely in the 1980s and 1990s, and has been documented in many impressive pictures, underpinning the reputation of XLPE cables as being water-sensible. However, many of these experiences were made with XLPE cables produced with steam-induced cross-linking, and extrusion of screens and insulation in subsequent operations, thus paving the road for water and dirt into the insulation. Worse still, in the first excitement on the new insulation material, some cables were laid directly into water with bare insulation. Today, all premium manufacturers produce XLPE cables with triple-extrusion and dry curing tubes. By these and other improving steps, the quality and breakdown voltage of XLPE insulation has improved dramatically since the 1980s. A comprehensive survey on water-tree phenomena and a large reference list for further reading can be found in [9]. Water-treeing is a problem not reported anymore. In contrary, XLPE has developed to be the insulation material of choice for both land and submarine power cables.

XLPE suppliers offer a number of cable materials with different properties. A well-composed formulation of base resin and additives can have excellent dielectric values, good processability, water-tree-retarding, and anti-aging properties. For a high-performance high-reliability insulation system, it is recommended to use a well-defined material combination of XLPE insulation material and semiconducting screen material recommended by an experienced cable manufacturer.

The dielectric properties of XLPE vary between different species and brands. The dielectric strength that is perhaps the most important property in cable design is not a constant fixed value. It depends not only on the material formulation but also on manufacturing conditions, sample preparation, temperature, test voltage shape, and much more. The concept of dielectric strength is based on a statistical distribution of experimental breakdown voltages on samples of a given insulation design (Weibull distribution). The details of the Weibull distribution of the dielectric strength are described in most cable textbooks and are needless to be repeated here. In a Cigré paper the application of Weibull distributions to cable insulation is described along with a comprehensive bibliography on the subject [10].

The use of XLPE cables for submarine application is limited by the availability of suitable joints. While XLPE cables per se are available for voltages up to 550 kV for land cables, there are no joints available for this voltage levels. While flexible joints have been made occasionally for 245–345 kV, higher voltages remain the realm of unjointed cables. This restricts the maximum reach of XLPE submarine cables to lengths, which can be produced in one piece. Submarine power cables for 170 kV a.c. can be produced with factory lengths of 50 km or more.

### *2.2.3 Conductor and Insulation Screen*

If we would extrude the XLPE insulation layer right away onto the conductor, the grooves, ridges and irregularities of the conductor would generate local stress enhancement and would reduce the dielectric strength of the insulation considerably.

To avoid this, a layer of semi-conductive XLPE is extruded onto the conductor resulting in an extremely smooth dielectric surface towards the insulating XLPE. As the semi-conductive inner layer is completely circular and has a smooth surface, there will be no stress enhancements.

The surface of the inner conductive layer can be made visible using hot-oil bath methods. XLPE becomes transparent at 130°C and reveals in detail the surface structure of the inner semi-conductive layer.

The triple-extrusion-method also provides a semi-conductive layer outside the insulation layer in order to form a stable dielectric surface not being affected by the outer screen layers. The three layers "conductor screen – insulation wall – insulation screen" make up the cable dielectric system. For high-quality insulation the three layers should always be manufactured simultaneously in a triple-extrusion system.

Semi-conductive XLPE-layers are made of PE-based co-polymers blended up with 40% carbon-black. According to international relevant standards, the volume resistivity shall be lower than 250 Ωm (CENELEC) or lower than 500 Ωm (AEIC, IEC). If the resistivity of the screen materials is too high, impulse voltages in the cable system can induce large stresses in the semi-conductive materials, eventually generating an insulation breakdown. Most modern carbon-black formulations meet the requirements with great margins. The resistivity of carbon-black filled XLPE in practical life is strongly dependent on temperature and mechanical stress, as the contact bridges between the carbon-black particles can close or open under the influence of strain in the polymeric matrix [11].

There are different varieties of carbon-black, such as furnace-black and acetylene-black with different particle size distribution and purity. Extruding parameters are strongly affected by the carbon-black content and character. Peelable screen systems with reduced adhesion between the insulation wall and the outer XLPE semi-conductive screen are sometimes used for medium-voltage cables. During the preparation of joints or terminations, the outer screen can be peeled off like a banana skin saving working hours.

The semi-conductive layers are applied with 1–2 mm nominal thickness. For long cables, the cost for a thick semi-conductive layer can add up to a considerable sum, as the raw material is fairly expensive.

### 2.2.4 The Influence of Ageing and Humidity on XLPE Insulation

"Ageing" is the deterioration of insulation material properties under the influence of temperature, electric and mechanical stress, chemical aggression, or any combination of these. Under ageing, the dielectric strength is decreasing gradually. When the dielectric strength has decreased to a degree that it is in the range of the operating stress the cable has reached the end of its useful life. Due to the statistical nature of the dielectric strength, some sections of a cable route will reach "pension age" and experience breakdown while the rest of the route still has years of service life. For statistical reasons, a longer cable route will face a larger risk of breakdown than a short route, everything else being equal. An analysis of the accumulated and expected repair costs helps the cable owner to decide when the economical end of useful life of the cable route will be reached.

Thermal ageing of insulation materials is described as a result of chemical reactions in the insulation. According to the classical Arrhenius model the life time can be expressed as:

$$\tau = A \cdot e^{(U/T)} \tag{2.1}$$

with: $\tau$: relative life time, $A$: constant, $U$: Activation energy for the chemical process in question, $T$: absolute temperature (K). Equation 2.1 can be set up for each of the chemical processes contributing to ageing when the activation energy and the Arrhenius constant are known.

Another equation has been established for the ageing as a result of electric stress:

$$\tau \cdot V^n = Const \tag{2.2}$$

with: $\tau$: relative life time, *Const*: constant, $V$: applied voltage, $n$: coefficient depending on material and manufacturing process. For XLPE cables $n$=9...12.

Neither Eq. 2.1 nor 2.2 deliver an exact calculation of the lifetime of a specified cable system. The value of the equations is the visualization of how the lifetime depends on the temperature and voltage. In most cases, the operating voltage of high-voltage cables is quite constant within narrow limits. It becomes obvious from the equations how much temporary overvoltages or overtemperature "eat up" the cable's lifetime. For certain cable installations (e.g. off-shore wind farms), it should be evaluated if there is the risk of long-time overvoltages in certain conditions.

For easy understanding and practical purposes, there are two fundamental rules for the lifetime of most organic insulation materials (polymers, paper, oil) [12]:

1. Increase the operating temperature by 8–10°C and you'll cut the life time by half.
2. Increase the operating voltage by 8–10% and you'll cut the life time by half.

Rule 1, known as the "Montsinger rule", has been proposed already in the 1930s. The rule can also be applied to modern insulation materials [13, 14]. According to this rule, a cable declared having a 30-years life at 90°C conductor temperature, would have a service life twice as long at 80–82°C conductor temperature. The 8–10° temperature step quoted in Rule 1 might have different values in other applications of insulation materials. For cable systems with varying load, a cumulative ageing calculation can take temporary thermal or electric stresses into account.

Many more ageing models have been proposed in the recent decades. Also the complicated process of multi-stress ageing under the simultaneous influence of thermal and electric ageing has been investigated thoroughly. References [13] and [15] provide a comprehensive compilation of this issue and list further references.

Copper has the potential to accelerate catalytically the ageing of polyolefinic (XLPE, PE) insulation materials. In modern cable designs the semi-conductive screen materials constitute an effective barrier for copper ions.

Humidity is the prime enemy of all electric insulation. It is a well-known fact that the dielectric strength and ageing resistance decrease under the influence of humidity. The insulation of the submarine cable must be protected from water and humidity by a purposeful design. It should be noted that polymeric materials are often completely water-tight but still allow for diffusion of humidity. If a complete humidity bloc is not possible for economic or other reasons, the water barrier should at least retard the water diffusion/intrusion so much that the humidity level inside the insulation stays under the critical limit during the economic life time of the cable. According to one strategy, the cable has a polymeric layer preventing direct water intrusion. Swelling tapes absorbs the humidity diffusing into the cable through the polymeric layer. The swelling tapes have sufficient capacity to absorb all humidity during the economic life of the cable [16].

Ageing by sun irradiation and UV can usually be excluded for submarine cables. The cable owner should provide a sun protection where the submarine cables come up to the surface on shore or on platforms/wind turbines.

### 2.2.5 Applications of XLPE Insulation

XLPE submarine cables are available for very long lengths (>50 km) in 3C fashion up to 170 kV from selected cable manufacturers. As voltage grows, the number of cable manufacturers is getting smaller. SC cables up to and including 500 kV can be manufactured in limited length. 400–500 kV flexible joints for XLPE cables are not available today (2009), which makes 400/500 kV extruded submarine cables difficult for extended sea routes.

### 2.2.6 Extruded HVDC Cables

Standard XLPE is not suitable for HVDC applications because of space charge phenomena in the insulation. Under the influence of a direct voltage, space charges would accumulate at certain places in the insulation wall. These accumulations would create unfavourable peaks of the electric field in the insulation. This phenomenon had rendered XLPE unsuitable for HVDC for a long time.

Today, special XLPE formulations have been developed that can cope with the problem. One manufacturer has gained more than 15 years of commercial experience with extruded HVDC cables. The first submarine extruded HVDC cable ("Cross Sound Cable") was installed in 2002 between Long Island, New York, and Connecticut in the US [17]. Since then, other extruded HVDC submarine cables from one supplier have been installed ("Troll A", "Estlink", "NordE.ON"). Extruded HVDC cable systems are now offered for 320 kV operating voltage. Given the

**Fig. 2.4** Submarine HVDC cable with extruded insulation, used for the "Estlink" link (Courtesy of ABB, Sweden)

conditions of submarine installation and operation, the mechanical sturdiness of extruded power cables is advantageous. Figure 2.4 illustrates a typical submarine HVDC power cable with extruded insulation.

### *2.2.7 Other Extruded Insulation Systems*

Ethylene propylene rubber (EPR) is an extruded dielectric used by some manufacturers mainly for underground cables.

Compared to XLPE, EPR has a rather high tanδ and $\varepsilon_r$ making it less suitable for the highest voltages. At least three manufacturers offer EPR medium-voltage submarine cables today. The latest known high-voltage submarine installation of EPR is the Venezia-Murano-Mestre 150 kV cable in Italy (2001). A comprehensive comparison of EPR vs. XLPE and relevant references can be found in [18].

An extruded PE d.c. cable was included in the Cross Channel HVDC installation already in 1963 [19].

### *2.2.8 Paper-Insulated Oil-Filled Cables for a.c. or d.c.*

This classical cable design is filled with low-viscosity oil. The cables are known under several names:

LPOF, low-pressure oil-filled,
SCFF, self-contained fluid-filled,
SCOF, self-contained oil-filled,

and possibly even more. There are also high-pressure type oil filled cables but these are not used for submarine applications. Mass-impregnated paper-insulated cables must not be confused with oil-filled cables. They are treated in the next section.

The insulation paper of oil-filled cables is made from conifer pulp (Kraft paper). To improve the dielectric losses especially for ehv a.c. cables, the pulp is sometimes washed with de-ionized water. Low-density paper (0.7–0.8 kg/dm$^3$) is being used to keep the dielectric losses low and the permeability for oil flow high. The insulation is often composed of paper tapes with different thicknesses, from 50 up to 180 μm. Thin paper, having a higher dielectric strength than thicker paper, is used close to the conductor where the electric stress is higher. In the outer parts of the insulation, the stress is lower and the required insulation thickness can be achieved with less layers of thicker paper tapes. An insulation wall with stacked paper thickness also provides good bending properties.

For decades the impregnation compounds for fluid-filled cables have been low-viscosity mineral oils. Since the early 1980s, synthetic cable fluids have been used in fluid-filled cables, predominantly linear alkylbenzene mixtures (LAB). As LAB is compatible with mineral oils it is also suitable for refilling and refurbishment of existing oil-filled cables.

During operation, the oil is pressurized. The dielectric strength of paper-oil insulation for a.c. voltage is depending on the pressure. The solubility of gases in the oil decreases with deceasing pressure. If the pressure falls too low, gases dissolved in the oil may be released and form bubbles. The electric field can create partial discharges in the bubbles, which can erode the insulation and cause an electric breakdown.

The necessary pressure in submarine oil-filled cables is maintained from pressure feeding units on the shore station. The cable must have sufficiently large oil channels to provide hydraulic communication from the land-based feeding units to all parts of the cable for the following cases:

– Thermal expansion and contraction. Responding to load changes the cable can change its temperature. The oil contracts/expands and the pressure variation must be equalized by oil transport to/from the land feeding units. Due to the flow resistance, there is a pressure drop from the centre of the cable to the land ends. The pressure variations are the largest in the middle of the cable route. It must be avoided at all costs that vacuum-like low-pressure conditions arise as these may cause electric breakdown.
– Positive overpressure against the seawater must be maintained at all locations to prevent water ingress in case of a damage of the metal sheath. Seawater is heavier than oil so that the pressure balance changes in favour to the water pressure as the laying depth increases. This must be compensated for by a larger oil pressure in the land-based feeding units.

The oil channel can be provided by different measures depending on the cable design. Single-core paper-oil cables have hollow conductors providing pressure equalization and hydraulic communication to the on-shore feeding units. In vintage cable designs the centre channel has been assisted by longitudinal grooves in the inside of the lead sheath. In three-core cables with common lead sheath, the interstices between the cores form oil channels. Today, only single-core oil filled cables are used for higher voltages (>150 kV). These rely on oil channels in the conductor centre.

Practical considerations show that the maximum length of fluid-filled cables is 30–60 km. For longer length, a sufficient oil flow in the cable can not be guaranteed. Only a lower coefficient of thermal expansion of the fluid, or a lower operating temperature would change this. The hydraulic design is discussed in detail e.g. in [20].

The Möllerhöj-cable is an exception to the rule of limited length of LPOF submarine cables (cf. Fig. 2.5). Comprising two or three fully insulated cable cores in a flat, untwisted formation, the thermal expansion of the oil can be absorbed by a slight deformation of the oval lead sheath which is supported by an intricate metallic structure acting as a spring-loaded cushion [21]. Möllerhöj submarine cables do not need external oil feed and can be installed in seemingly unlimited lengths. The latest submarine cable of this kind was the Kontek link between Denmark and Germany with 600 MW d.c. power in 1996. The cable will be replaced by a conventional mass-impregnated HVDC cable in 2010.

Oil-filled cables could be improved by using paper strips reinforced with polymeric films. Laminates composed of paper layers and polypropylene (PP) layers have been used as insulation matrix for oil-filled cables since the 1980s for voltages up to 800 kV a.c. (cf. Fig. 2.6). The material has a better dielectric strength for a.c.

**Fig. 2.5** Möllerhöj-type cable for the Kontek HVDC link (Courtesy of NKT Cable A/S, Denmark)

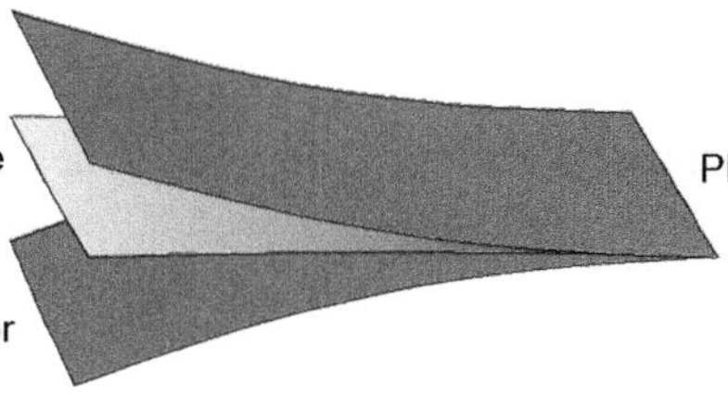


**Fig. 2.6** Layers of PPLP

and lower losses than Kraft paper which makes it attractive for high-performance EHV a.c. cables. The basic principle has been used with different sandwich schemes (paper-PP-paper, PP-paper-PP) in double, triple and quadruple layers. The insulation is known under the names PPL, PPP, or PPLP. More information and references on the development of PPLP cables can be found in [6].

Fluid-filled cables have a longer operational record for EHV a.c. cable links compared to XLPE cables. A well-proven range of joints and cable terminations is available. The high conductor temperature of 90°C allows for large ampacities, and the low losses of PPLP make longer cable lengths possible. On the negative side, there is the rather complicated production process, the complicated feeding stations, and, more important, the doubtful environmental behaviour with outpouring oil in case of a submarine damage. The positive properties resulted in the cable type still being used successfully for shorter submarine cable routes with high transmission power demands. The Strait of Gibraltar cable system, the Hainan-China Mainland Strait cable system, and the Jordan-Egypt (Gulf of Aqaba) connection are modern examples of a technique that has been used worldwide for distances of up to 40–50 km. Today, XLPE cables are qualified for very high voltages and leave little room for fluid-filled cables. Still, a system with four oil-filled submarine cables with only 138 kV rated voltage was installed in the Philippines by a Japanese manufacturer as late as 2000, although oil-related accessories such as feeder tanks and pumps could not even be transported to site by road [22].

### *2.2.9 Paper-Mass Insulation for HVDC*

This insulation type has been used for HVDC cables since more than 100 years [23]. Mass-impregnated cables have been used for medium-voltage a.c. transmission, but today, they are only being used for submarine d.c. applications for large transmission power at the highest d.c. voltage. Mass-impregnated cables are available for up to 500 kV d.c. For this voltage, there is no alternative to the well-proven mass-impregnated cables. The longest route length today is 580 km (NorNed, 2008) but this type of cables can be used for virtually infinitive lengths. As cables with paper-mass-insulation have many similarities with cables with extruded insulation, they are also called "solid-insulation cables". Figure 2.7 shows an 450 kV HVDC cable with mass-impregnated insulation, in comparison to a 150 kV cable from 1954.

**Fig. 2.7** Mass-impregnated HVDC cables: 450 kV "Baltic Cable", left, and 150 kV "Gotland cable" from 1954 (Courtesy ABB, Sweden)

Mass-impregnated d.c. cables require different insulation paper than oil-filled a.c. cables. As d.c. cables are not subject to dielectric losses, one can chose high-density paper ($\approx$1.0 kg/dm$^3$) to achieve best possible dielectric strength. Specialized paper makers are able to keep the porosity high in order to enable lower flow resistance through the papers. The paper used for submarine HVDC cables is made from conifer cellulose pulp (Kraft paper). Sometimes manufacturers use different paper tape thickness in the insulation to accomplish a better mechanical flexibility. For high-performance cables, the lapping process must be done under controlled humidity and requires scrupulous cleanliness.

The conductor screen in paper-insulated cables is made from semi-conducting carbon-black paper. The insulation screen consists of carbon-black paper combined with metal-laminated papers or thin-wire copper nets. Sometimes also duplex-paper (one side insulating, the other side carbon black) is used as screen. Both screens provide a smooth dielectric surface towards the paper insulation. Metal-laminated paper is also known as "Höchstädter Papier" since its introduction in 1914.

After lapping of all paper tapes, the cable core is put into giant vessels for further treatment. A vacuum-heat-drying phase is followed by hot impregnation with high-voltage grade impregnation compound. One certain mineral-oil based compound with non-Newtonian rheological behaviour has established itself as material of choice with European mass-impregnated cable manufacturers. The high viscosity is achieved by poly-isobutylene additives.

Mass-impregnated HVDC cables with high-viscosity compounds can be used for indefinite route length because they are not depending on external pressurization from on-shore feeding stations [27].

Mass-impregnated cables have small voids in the butt gaps of the insulation when the insulation is cold. Under the influence of the electric field partial discharges might occur in the voids. Under a.c. voltages the voids would ignite in every half-cycle. Many repeated partial discharges at the same location can lead to paper disintegration and eventually to breakdown. For this reason mass-impregnated cables cannot be used for high voltage a.c. In a d.c. field a different charge transport mechanism allows for partial discharge to ignite only very rarely. When the cable is getting warm due to conductor current the impregnation compound expands filling all possible voids. The dielectric strength of a warm mass-impregnated cable is much higher than that of a cold cable.

Thanks to the layered structure, the mass-impregnated insulation can tolerate small impurities and/or defects. Scattered defects and foreign particles smaller than the thickness of a paper layer, can be regarded as harmless as long as they do not emit ions.

Attempts to replace the paper in mass-impregnated cables by paper-propylene-laminate have been made. One manufacturer using PPLP together with the standard impregnation compound (with altered viscosity) reported no significant improvements compared to standard paper-mass insulation [24]. A more recent publication by another manufacturer reported better power rating but this seems to be achieved by using low-viscosity impregnation oil and a higher current rating [25] which resembles an LPOF cable. The use of PPLP for mass-impregnated dc cables seems to require further development. Furthermore, the authors suggest the change of relevant test standards before PPLP can unfold its possible advantages.

Cables with mass-impregnated insulation do not leak oil to the environment when they get damaged [26]. This property is important for submarine cables to be installed in sensitive environments.

### 2.2.10 Gas-Filled Submarine Cables

Lapped cable insulation with a gas filling rather than an oil filling was developed by C. J. Beaver and E. L. Davey of W. T. Glover & Co already in 1937. Having the same or similar design elements as a standard oil-filled cable, the insulation was constructed from pre-impregnated paper tapes. After installation the cable was vacuum-treated and pressurized with nitrogen from the cable ends. The pressurized gas filled the band gaps between the paper tapes suppressing the formation of partial discharges. The gas filled cable can be used for a.c. or d.c. voltages despite the voids in the insulation.

Submarine 138 kV single-core gas-filled cables with gas-filled insulation were installed between British Columbia and Vancouver Island in 1956–1958 (Fig. 2.8). One system has been removed in 2007, and the other is expected to serve ten more years [28].

Three gas-filled submarine cables for an HVDC scheme (one spare cable) were laid under the 40 km wide Cook Strait between the two islands of New Zealand in 1962.

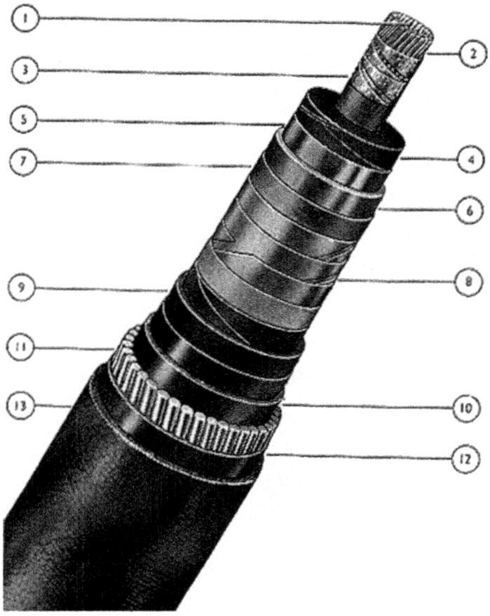

1. Gas channel
2. Copper conductor
3. Conductor screen
4. Paper insulation
5. Dielectric screen
6. Lead-alloy screen
7. Bedding
8. Reinforcement
9. Reinforcement protection
10. Bedding
11. Steel wire armoring
12. Cotton tapes
13. Serving

**Fig. 2.8** Vancouver Island 138 kV gas-filled cable [28]

### *2.2.11 Other Insulation Systems*

Since the first days of cable industry all kinds of material were tested for their suitability as insulation material. In the beginning, fibrous vegetable products such as cotton, jute, flax and hemp were examined. Very early, it was detected that hydrophobic impregnation compounds could improve dielectric or breakdown properties. Mineral or vegetable oils, their combination, derivatives and by-products all were tested in great varieties. The Scottish physician William Montgomery exhibited samples of a Malaya plant sap called gutta-percha at a meeting of the London Royal Society of Art in 1843. In 1845, the samples found William Siemens's interest, and he sent them to his brother Werner in Berlin for further evaluation. The material was extrudable and served as cable insulation for at least 3 decades. The next widely used insulation material was vulcanized rubber. These insulation materials had an important roll in the development of submarine cables, but today, they do not even "surface" in repair cases.

In modern days insulation systems employing nano-systems are being considered. Cigré has prepared a summary report on this item [29], providing comprehensive further readings. Another more recent update can be found in [30]. In nano-insulation materials, the insulating matrix is enriched with inorganic nanoparticles. Some insulation material properties such as pd-resistance, space-charge properties, water-tree-resistance etc, could be improved. Before a deployment in the field of submarine cables will be possible, this concept will have to prove itself in underground applications ashore.

Gas-insulated lines (GIL) have been installed in a few occasions for the transmission of very high power values onshore. They consist of a conductor hold in place by polymeric spacers inside a metallic tube. The space between the central conductor and the tube is filled with pressurized insulation gas such as $SF_6$, nitrogen or a mixture of these gases. Gas insulation of this type has a long successful record for switchgear installations. Designed for voltages of 245 kV and above, GIL can transmit enormous power. Gas insulation relies on extreme cleanliness of the gas volume. There are a few onshore GIL installations already, comprising a length of a few hundred meters each. GIL have been suggested for submarine links to transmit the enormous power flow from planned offshore wind farms in Northern Germany. Although the installation of welded pipelines is a well-known technology, a submarine application of GIL would require the mastering of extreme cleanliness during the assembly on-board. A possible ingress of seawater in case of a leakage would contaminate an enormous length of the GIL requiring a complete new installation. It remains to see if this type of transmission can be installed and operated safely in the harsh submarine environment.

## 2.3 The Water-Blocking Sheath

The dielectric insulation must be protected against undue water ingression to maintain the dielectric strength. Most high-voltage submarine cables have a metallic sheath to provide protection against water ingression. Aluminium, lead, copper, and other metals can be used for this purpose in a variety of shapes. Medium-voltage cables often can do without metallic sheaths or with a simpler design of sheath because of the low electric stress in the insulation. One strategy is to equip the cable with some water-absorbing agent under a polymeric sheath. The polymeric sheath would be watertight but would allow little humidity to diffuse through the sheath in the vapour phase. The absorbing agent has sufficient capacity to keep the insulation dry enough for the entire economic life of the cable and beyond [16].

Most metallic sheaths also provide a good protection against the Teredo, a family of aggressive "shipworms" (actually, they are mussels) feasting not only on sunken pirate ships but also on submarine cables. It is abundant in most saltwaters. In recent years there are indications that the Teredo has invaded also the brackish Baltic Sea as an alien species. They can be kept at bay also by copper or brass tapes.

### *2.3.1 Lead Sheath*

Wheatstone and Cooke suggested lead sheaths for telegraph cables already in 1845. Lead extrusion was known since 1797, but it took the relentless efforts of some known and countless unknown cable engineers to arrive at the lead sheath quality of today. Before 1900, most of the large cable manufacturers were using a ram press to produce lead-covered cables [31]. Well-produced lead sheaths are completely impermeable for water ingression and humidity diffusion. For oil and mass-impregnated cables, they provide an enclosure to protect the environment outside the cable. Furthermore, the lead adds to the cable weight, which in certain cases is important for the stability of the cable on the sea floor. With screw extruders, submarine cables of 100 km length can be lead-covered in a single uninterrupted length at reasonable costs. Long term stability, creep, and extrusion properties can be improved substantially by using lead alloys with alloy elements such as antimony, tin, copper, calcium, cadmium, tellurium, and others. The standard EN 50307 lists a number of lead alloys for cable use. Table 2.3 shows the composition of lead alloys, which have been used successfully for the extrusion of long submarine cable lengths. A longer table is given in Chap. 12.

Lead and lead alloys are very soft and must be protected against mechanical damages during manufacturing, cable transport, installation, and beyond. In early times, it happened far too often that the lead sheath was ripped open by misaligned cable rollers or protruding edges. Also, early lead sheathing machines (before extrusion, a discontinuous ram press was often used) delivered sheaths with wrinkles, air intrusions, holes, etc. The risk of damaged lead sheaths was so high that many cables were produced with a double lead sheath.

Modern submarine cables are often lead-covered in extruders with high reliability providing lead sheaths with narrow thickness tolerances and good alloy stability. When screw-extruders are used for lead covering, a PE extruder can apply a PE sheath immediately after the lead extruder, thus protecting the lead during the subsequent manufacturing. On-line thickness measurements of the lead sheath and the plastic sheath contribute to a high-quality sheathing system.

The lead sheath is subject to fatigue processes. Vibrations, repeated bends, and thermal cycling can result in a recrystallisation of the lead alloy. Crystal boundaries can develop to micro-cracks and deteriorate the water-tightness of the lead sheath. Wave induced movements of the cable-laying vessel can subject the cable that hangs down from the vessel to many repeated bendings. Depending on wave characteristics, amplitude, number of bends, and bend radius, this might cause premature fatigue in the lead sheath. Eventually, the lead sheath can break.

In order to mitigate risks of fatigue due to thermal cycling, lead-covered cables have been used with an oval conductor-insulation-lead sheath system (e.g. SACOI

**Table 2.3** Lead alloys and their constituents suitable for submarine cables. The table shows the designation according to EN 50307, and the conventional names for the nearest related alloy

| Alloy designation acc. to | | Alloy elements and percentage (by weight). Min and max values | | | | |
|---|---|---|---|---|---|---|
| EN50307 | Convention | As | Bi | Cd | Sb | Sn |
| PK012S | ½C | | | 0.06–0.09 | | 0.17–0.23 |
| PK021S | E | | | | 0.15–0.25 | 0.35–0.45 |
| PK022S | EL | | | | 0.06–0.10 | 0.35–0.45 |
| PK031S | F3 | 0.15–0.18 | 0.08–0.12 | | | 0.10–0.13 |

1967 200 kV d.c. cable, Vancouver Island 138 kV gas-filled cable [28]). The idea is that thermal expansion of the cable would result in a deformation from an oval shape into a circular shape of the cable rather than a strain and re-compression of the lead sheath.

### 2.3.2 Aluminium Sheath

Aluminium sheaths come in different shapes: extruded, welded, or laminated.

For the welded sheath, aluminium strips in the range of 0.5 up to 4 mm thickness are folded around the cable with a set-up of rolls. The strip edges are trimmed to correct dimensions and welded longitudinally forming a tube. The welding can be made with laser welding methods. The tube-like sheath can be corrugated after welding to improve the flexibility. The weld seam must be checked carefully for pinholes. Welded corrugated cable sheaths can be made from more corrosion resistant alloys than extruded.

The extruded aluminium sheath has been employed for many decades in straight and corrugated shapes. Extruded aluminium sheaths in a thickness of 2 up to 4 mm have been used occasionally for submarine cables, but heavy corrosion problems occurred. Today, extruded aluminium sheaths are not used for submarine cables anymore.

Laminated aluminium sheaths consist of a thin aluminium foil (0.1–0.3 mm thick) pre-laminated with a layer of PE-copolymers. During the cable production, this laminate is formed around the cable core with the polymeric layer outside. The edges of the laminate strip are glued together with overlap. In the same operation, a PE sheath is extruded directly onto the PE layer of the aluminium laminate. The laminate is thus firmly bonded to the outer PE sheath. Only by this bonding, the aluminium sheath is able to withstand the cable bending without wrinkles or creases. In most cases, there is a copper wire screen directly under the laminate to carry the short-circuit current. Laminated aluminium sheaths have an excellent track record from land cables. While the aluminium foil is absolutely impermeable for water and humidity, tiny amounts of humidity might diffuse into the cable through the glued seam. The amount is much depending on the seam geometry and the used materials. Z-folds can increase the diffusion path and bring down the diffusion rate considerably. Water adsorbing agents inside the cable keep the humidity level down to an acceptable level for the safe use of aluminium laminate cables in submarine medium-voltage applications.

### 2.3.3 Copper Sheath

Copper sheaths from welded and corrugated copper strips are sometimes used in submarine power cable applications. The corrugation machine can provide a wave structure with annular or helical gaps. For submarine cables, the annular gaps structure is preferred because it provides a better barrier against longitudinal water migration in case of damage. The profile of the corrugation "waves" may be sinusoidal, trapezoid, or another shape. These shapes have different properties regarding internal or external pressure resistance, and bending and fatigue properties. Copper sheaths are corrosion resistant and, at proper sizes, even able to carry short-circuit currents. Separate copper wire screens are dispensable. Copper properties can be improved by using alloy elements such as beryllium.

A copper sheath is considered to be very resistant against fatigue phenomena, which can demolish a lead sheath after a large number of bends. For this reason, a copper sheath can be used for dynamic power cables, which are suspended freely from floating oil and gas platforms and are subject to repeated bends due to waves.

### *2.3.4 Polymeric Sheaths*

Polymeric cable sheaths have different functions depending on where they are located in the cable construction. A polymeric sheath usually protects the underlying lead sheath from corrosion and abrasion. For this purpose, mostly HDPE or LDPE sheaths are used. They provide excellent chemical and mechanical stability over a long life at moderate costs. Other materials used for sheathing are PVC, polyamide ("Nylon"), and polyurethane. Polyamide has even better mechanical properties than HDPE.

If there is no metallic sheath, a polymeric sheath can provide a radial water barrier. Polymeric materials are water impermeable, but water vapour can diffuse through them. For medium-voltage submarine cables, the small amount of water vapour diffusing through the polymeric sheath can be accepted as the electric field strength in medium voltage cables is small. The most common cable-sheathing polymer, HDPE, has a rather low water vapour permeability (145 $g\mu m/(m^2 \cdot d)$, @38°C, 90% r.h., according to a resin supplier). Both PVC and polyamide (Nylon) have higher diffusion rates. Specialty polymers such as PVDC have much lower diffusion rates than HDPE but have not been used for power cable sheathing yet.

Small scratches or even small holes of the plastic sheath do normally not constitute any danger for the corrosion protection, as the access of oxygen-rich water is very limited.

Polymeric sheaths are sometimes made from semi-conducting PE materials loaded with carbon-black. They provide voltage equalization between the metallic sheath inside and the armoring outside.

## 2.4 Armoring

The most prominent construction element of submarine cables is the armoring, which provides both tension stability and mechanical protection.

For every submarine cable project the armoring should be designed with respect to the tension stability, external threat pattern and protection requirements for each sector of the planned cable route.

Submarine cables are exposed to tensional forces during the installation not only by the weight of the hanging cable, but also to additional dynamic forces from the vertical movements of the vessel. The total force during installation can dramatically exceed the static force of a cable hanging down to the seafloor. Tensional forces and tensional stabilities are discussed in Chap. 3.

The armoring must also provide sufficient mechanical protection against expected external aggression by installation tools, fishing gear, and anchors.

Submarine power cable armoring is built from metal wires wound around the cable with a certain lay length (also called pitch). The lay length, i.e. the length of cable in which the armoring wire completes one turn around the cable, is between 10 and 30 times the cable diameter under the armoring. Round armoring wires have a diameter of 2–8 mm.

The design of the armoring has a strong influence on the cable properties such as bending stiffness, tensional stability, torsion balance, and on the choice of handling and installation methods. Some relations are given in Table 2.4.

A helical armoring translates a tensional force into a torsional force trying to twist the cable. In a long lay-length armoring, the wires run almost parallel to the cable axis and can take up tensional forces without building up too much of torsional forces. The tensional stability of the cable is very large. At the same time, a long lay-length increases the bending stiffness of the cable, which is undesirable.

Submarine cables with short lay-length in the armoring have generally a lower bending stiffness. Subject to tensional forces, the short lay-length armoring is tightened up inwards without taking up much of the tensional force. Much of the force is transferred to the conductor instead.

**Table 2.4** Properties of some armoring concepts

| Properties of armoring of single-core cables with single-layer armoring | | |
|---|---|---|
| | Long lay-length (pitch) | Short lay-length (pitch) |
| Tensional stability | ++ | 0 |
| Bending stiffness | – | + |
| Torsional stiffness | + | 0 |
| Possibility to coil the cable | ++ | + |

| Properties of armoring of single-core cables with double-layer armoring | | | | |
|---|---|---|---|---|
| | Unidirectional lay | | Counter-helical lay | |
| | Long lay-length (pitch) | Short lay-length (pitch) | Long lay-length (pitch) | Short lay-length (pitch) |
| Tensional stability | + | – | ++ | o |
| Bending stiffness | – | + | – | + |
| Possibility to coil the cable | + | 0 | – | – |

The armoring lay-length must be optimized with respect to the expected tensional forces, the tension stability of the conductor and the torsional requirements of the cable and its installation.

A cable with a unidirectional armoring can absorb torsions in one direction only: the direction where the armoring layer is opening up. This allows laying up the cable by coiling it on a storing area or fixed cable tank.

A double layer of armoring wires (DWA) provides a much stronger protection against external force than a single layer. When the two layers have different lay direction, they can prevent intrusion of sharp edges from anchor flukes, cable ploughs, rocks, etc. In a counter-helical DWA, the torsional forces of each layer counterbalance or even cancel out each other. Therefore, submarine power cables for large depths of water are equipped with torque-balanced counter-helical armoring layers.

A special combination of armoring layers is called "rock armor" and comprises an outer wire layer with very short lay length (cf. Fig. 2.9). This layer with short lay does not contribute to the tensional strength but improves significantly the crush resistance properties of the cable and results in a lower bending stiffness than a long-lay armoring [11]. This design gives a better protection against external damage caused by impacts from rocks, dropped objects and towed equipment such as fishing gear.

Counter-helical armored cables require turntables for storage and installation. With the two layers having the same lay-up direction, the cable still can be twisted and coiled. However, the necessary twisting forces can become very high. During

**Fig. 2.9** Three-phase submarine cable with double-layer round wire rock armoring (Courtesy of ABB, Sweden)

laying, high tensional forces in connection with unidirectional lay create considerable torsional forces.

Flat armoring wires rather than round wires have been used successfully. They provide the necessary steel cross section with less outer diameter, saving some material and space on the laying vessel. Flat wires are more expensive than round wires. In early times of cable manufacturing, even slot-and-key wires have been used for the armoring. The wires interlock and provide a closed yet flexible armoring where no single wire can be removed [32].

For shallow water with low tensional requirements, open armoring with spaced wires can be used. The spaces can be left open or filled with spacer wires made from plastics, ropes, or something similar. Open armoring not only reduces weight but also, in case of a.c. cables, can reduce the eddy current losses. Open armoring is used but seldom as it offers less mechanical protection than a closed armoring.

In most cases, the armoring is made from mild steel, which is a magnetic material, concentrates the magnetic field around the conductor, and generates unwanted losses and extra heat in a.c. submarine cables. In single-core a.c. submarine cables, the losses in the steel armoring can reduce the ampacity substantially. Some strategies have been developed to reduce these losses:

1. Armoring made from non-magnetic materials, such as bronze, brass, copper or aluminium. Table 2.5 lists the tensile strength of some non-ferrous metals. Copper-based alloys are an expensive choice for an armoring. Aluminium is much cheaper but can easily be attacked by seawater corrosion. Armoring made from copper wires combines low resistivity with good corrosion resistance, but the mechanical strength is less than steel. Hard-drawn copper wires have a substantially better mechanical strength but slightly lower conductivity than annealed copper. A double counter-helical flat wire copper armoring has been used down to 400 m for the 500 kV a.c. submarine cables between British Columbia and Vancouver Island in the early 1980s [33].
   A superior yet expensive solution is the use of stainless steel wires for the armoring. They provide a low-loss non-magnetic armoring with high tensile strength and very good seawater corrosion resistance (valid for stainless steel with high alloy element content).

**Table 2.5** Composition and tensile strength of non-ferrous armoring materials

| Material | Designation | Tensile strength (MPa) | Composition |
|---|---|---|---|
| Copper | Cu–ETP | 200–400 | Cu 99.9% |
| Brass | CuZn40 | 340–500 | Cu 59.5–61.5%, Zn: remainder |
| Phosphor bronze | CuSn6 | 340–1000 | Sn: 5.5–7.0%, Zn: a few % Cu: remainder |

2. Reduction of magnetic induction. The single-core a.c. cable can be provided with a massive copper screen, which is firmly grounded on both ends of the cable. A screen current is generated almost as large as the conductor current. These two currents run in opposite directions inside the armoring and virtually cancel their magnetic fields. The magnetic losses vanish almost completely but losses in the copper screen will arise instead. To achieve this goal the copper screen needs to have a cross section similar to the conductor. A counter-helical double armoring, the inner layer being of copper, the outer of steel, can make up a low-loss, high-strength armoring for single-core a.c. cables.

In three-core-cables, the magnetic fields of the three cores cancel each other to a great extend, and the magnetic losses are reduced to a lower level.

The tensile strength of steel armoring can also be achieved with modern aramid or polyester fibres (trade names: Kevlar, Vectran, Twaron, etc). Lightweight aramid fibres have been used to reinforce special submarine cables for the offshore industry since the 1970s [34]. Cords or wires made from these fibres provide only little protection against side impacts such as anchoring or fishing gear aggression. One manufacturer developed an armoring based on fibreglass-reinforced plastic (FRP) in 1985 and installed a three-core cable with one layer of polymer-sheathed FRP threads and one conventional steel wire layer between two islands in the Japanese Okinawa prefecture in 2000 [35]. The creep behaviour of ultra-high molecular-weight polyethylene fibres (trade names *Spectra* or *Dyneema*) render them not being useful for submarine cables under sustained load [36].

The armoring is applied on a bedding which protects the underlying surface (often an extruded inner sheath) from undue located press from the armoring wires. Beddings are made from fabric tapes, which offer a certain amount of compressibility. A special type of bedding tapes is known under the name Hessian tape.

### *2.4.1 Corrosion Protection*

Marine salt water is a corrosive environment. In the open oceans the salt content is 33 through 39‰ by weight. In marginal seas, inland seas and coastal areas the salt content can vary considerably due to freshwater admixture from rivers, or evaporation under solar irradiation. The open North Sea has a salinity of 32 up to 36‰, decreasing to 15 up to 25‰ close to shore. The Baltic Sea holds 3 up to 25‰. In tropical waters the local salt content can vary largely due to strong evaporation and/or large amounts of tropical rain. Salinity also varies through the annual seasons, and with depth. Sometimes the unit *psu* (Practical Salinity Unit) is being used, with 35 $psu = 35‰$ by weight.

Corrosion must be considered seriously when designing submarine power cables. There are several corrosion mechanisms each deserving close attendance. General corrosion occurs by the impact of salt seawater onto the armoring wires. The armoring wires are in most cases made from zinc-coated steel wires. A zinc layer of 50 μm or more is the primary corrosion protection of steel wires. The secondary protection is made by flushing the armoring with hot bitumen during manufacturing. The lifetime of the bitumen protection is much depending on the mechanical impacts on the cable. The bitumen layer can be eroded during installation or later during service when the unburied cable is hit by water currents laden with sand. Where the bitumen layer is damaged, the zinc layer takes over the corrosion protection. The corrosion rate of zinc coatings in seawater is depending on many factors such as salinity, temperature, water exchange around the cable, etc. Reference values of the corrosion rate of zink layers are 5–50 μm/year [37, 38]. Investigations of retrieved cables after decades of exposure in the Baltic Sea have shown a zinc layer corrosion rate of about 1 μm/year. When the zinc is gone, the steel may be consumed at rates of 10 μm/year [39]. An important role for the corrosion plays the free access to fresh sea water containing dissolved oxygen. If the water exchange around the cable is inhibited by soil or protecting layers, the existing oxygen close to the cable is not replenished and the corrosion rate will decrease considerably.

Electric corrosion has been identified as a possible strong effect. The corrosion is associated to electric currents entering and leaving the steel wires from the ambient. Tidal flows in combination with the earth's magnetic field are alleged to cause these currents [35]. Also stray currents from onshore railways, nearby cathodic protection systems of pipelines, or strong welding operations have been identified as sources of electric corrosion [39].

Sacrificial zinc wires are sometimes used to protect steel wires from corrosion, much similar to the anti-corrosion zinc anodes used in naval applications. This anti-corrosion protection method works only if the sacrificial zinc anodes have a sufficient good galvanic contact to the protected steel wires. Given the bitumen flushing during the manufacturing process, the electric contact between the wires is only random and the zinc wire method is not considered to be very successful.

Also semiconductive polyethylene can be involved into electric corrosion. It may act as a galvanic agent connecting two metals with different electrochemical potential leaving the lower rated metal to corrosion. Also, the semiconductive polyethylene can suffer from decarbonisation and eventually punctuation as a result of a d.c. stray current over a certain density [39].

The armoring wires can also be corrosion-protected by an individual polymeric sheath. The direct contact to salty seawater is excluded, as long as the sheath is not washed away by sand abrasion. The method can avoid much of the troublesome bitumen layer. The method has been used occasionally and might be considered for OWP in-field cables. When preparing cable ends for terminations on offshore wind turbines, the brittle bitumen will crack from the cable ends and soil down the place. Here, plastic-coated armoring wires could be a better alternative. However, when damaged, the individually extruded sheaths could lead to an adverse effect, when pitting corrosion occurs from localised current going into the wire.

Corrosion affects but a fraction of armoring wires at singular spots. The remaining wires still offer a protection against external forces even if the tensile force of the cable has decreased.

The armoring can also be constructed from more corrosion-resistant metals. Copper, bronze, and brass wires have been used as armoring. However, aluminium should not be used in seawater. The 1970 Long Island cables suffered severe seawater corrosion of its armoring made of Al–Mg–Si "Aldrey" alloy [40]. Aluminium armoring has been used in brackish waters [41] with doubtful results, but can be considered for freshwater.

The use of an outer extruded plastic sheath over the complete cable as a corrosion protection is useful as long as the sheath is undamaged. Local damages in the sheath can concentrate corrosion-enhancing galvanic currents.

## 2.5 Outer Serving

Scratches can deteriorate the anti-corrosion effect of bitumen and zinc layers. To avoid this, an outer serving protects the corrosion protection of the cable armoring during loading, laying, and burying of the submarine cable. Modern submarine power cables have either extruded polymeric outer servings or servings made from wound yarn layers. Cables with wound yarn layers are designed for the seawater penetrating into the armoring wires to the plastic inner serving (semi-wet design). The water exchange in the narrow interstices under the yarn layers is very limited and reduces corrosion rates considerably. The yarn layers play a sacrificial role, and small damages during handling or laying are not considered harmful. For power cables intended to be coiled for storage or installation, the lay direction of the yarns should be the same as for the armoring wires. Otherwise the yarn layers can easily be exploded by the opening of the underlying steel wires.

An extruded outer serving provides a neat appearance of the submarine power cable and most often a dry environment for the armoring wires. Still, the armoring system must be designed for water exposure in case of damage in the extruded serving. The damage could soak the complete armoring with seawater and enhance local corrosive stray currents.

The friction coefficient of the outer serving must be taken into account for the design of the installation procedure. Wound yarn layers provide a good grip for cable pulling engines on board the CLV, while extruded servings are more slippery.

The outer serving usually has markings to make the cable visible for ROV cameras and to distinguish the cable from other cables on the seafloor. Extruded outer servings can be provided with a longitudinal strip in a different colour. In wound yarn servings, some of the outer yarn threads are usually replaced with threads of different colours generating a helical pattern. White, yellow and orange strips provide a good contrast to the black background colour. When laying more than one cable on a confined space, a different marking of the cables can help to identify the cables on the seafloor. For a pair of HVDC cables, it makes sense to identify one cable with one stripe, the other with two.

The inner serving layer of submarine cables is often soaked with bitumen in order to corrosion-protect the underlying steel armoring. The bitumen can cause some problems during the cable installation. In warm weather, the bitumen may exude from the cable on-board of the cable ship and aggregate on the cable rollers and cable engines to thick disturbing layers. Also, when preparing cable ends for termination on offshore installations, such as offshore wind turbines, the brittle bitumen will crack from the cable ends and soil down the place. For these cases, plastic-coated armoring wires could be a better alternative. The sticking of bitumen-soaked cables on the turntable can be relieved by coating the outer serving with lime wash.

## 2.6 Three-Core Cables

Modern medium-voltage submarine cables ($\leq$52 kV) are usually designed as three-core (3C) cables with XLPE insulation.[2] The cable cores are similar to those of land cables. Many different design alternatives are used for this class of submarine cables. Many cable constructive layers can be applied either to each single core, or to the three cores in common. Furthermore, many different screen/sheath configurations can be found.

Depending on the specifications submarine medium-voltage cables can be produced with or without a metallic sheath. The low electric stress in extruded medium-voltage cables allows for a cable design without total lock-out of water. An extruded plastic sheath provides a secure water barrier while water-absorbing agents inside the plastic sheath can take care of the minute amount of water vapour, which can diffuse through the extruded plastic sheath.

Another common system is the aluminium laminate for water protection combined with a copper wire screen. The aluminium laminate constitutes a radial water barrier, while the copper wire screen is designed to carry fault currents.

The non-metallic sheath of three-core submarine cables can be produced from semi-conducting polymeric materials. The capacitive currents of the three phases can balance each other, which can increase the cable ampacity. Many different combinations of sheath, screen, and corrosion protection are possible. Manufacturers have their own "recipes" to meet the very different conditions and requirements of submarine power links.

High-voltage cables are available in 3C design up to and including 170 kV. The first 245 kV 3C submarine cable has been installed in a moderate length of 7.8 km in Canada in 2008. This system voltage is possible for short cables without the requirement of flexible joints, which are not available yet for higher voltages. Almost all modern high-voltage 3C cables are produced with XLPE insulation.

Once the cable cores are manufactured and tested, they are laid-up to provide the flexibility of the cable. Beside standard S and Z lay-ups, there is also a S-Z-lay-up for mv cables, where the cable cores are twisted right-hand and left-hand in sequence. After lay-up, the diameter of the 3C cable over the core binder is 2.16 times the diameter of the individual cores. Four interstices are generated during the lay-up, three in the periphery and one central. To provide a stable circular base for the armoring, the outer interstices are normally filled with fillers. Ropes of different

[2]Some manufacturers still offer medium-voltage submarine cables with mass-impregnated insulation. They may have shaped individual cable cores providing a circular base for the lead extrusion.

sizes or tailor-made extruded polymeric profiles are used to fill the space. Extruded profiles from PE or PVC provide an excellent support for the armoring wires and can accommodate small optical cables. PE requires more complicated extruding dies but lower material costs per unit length of the cable, compared to PVC. For very long submarine cables, the use of PE is more economic, and for short length, PVC is the cheaper alternative. Even lead profiles have been used as filler profiles, when the cable needs additional weight for an installation in streaming water.

Filler ropes can be made from recycled polymeric material of any kind. A combination of three ropes of two sizes can fill the interstices reasonably well.

The central space of the cable usually does not need a filler for stability. The three cable cores support each other so they do not collapse into the central space. It is not suitable to include an optical cable into the central space because it might be exposed to tensional forces as the central line is not the neutral line in cable bending.

After lay-up, or sometimes in the same operation, the cable is fed into the armoring machine. In 3C a.c. cables, the magnetic fields from the conductor currents cancel each other to a large extent. Therefore, the armoring can be made from steel wires without excessive magnetic losses. 3C cables for high voltage can easily reach a diameter of 200 mm and more. As the number of spools in the armoring machine is limiting the number of armoring wires on the cable, sometimes very thick steel wires must be used to cover the entire circumference of the cable. For a cable with 200 mm diameter (under the armoring) and an armoring machine with a capacity for 84 wires, the combined steel cross section of a single armoring layer would be about 3700 mm$^2$. This massive armoring would be sufficient for more than 300 m laying depth. Flat armoring wires can offer sufficient steel cross section for shallow waters and still cover the complete cable circumference. However, the protective effect against external aggression of armoring wires goes down with the square of the wire diameter. Many cable operators think that the better protection of round wires justify the increased material costs compared to flat wires.

3C cables have been produced also with oil-filled insulation for up to at least 170 kV. These cables are always equipped with a lead sheath, either a common lead sheath for all three cores, or individual lead sheathes for each core. In a common lead sheath, the interstices between the cable cores provide conduits for the oil transport to and from the "breathing" cable under the influence of load changes. An oil conduit in the centre of the conductor is not necessary. Filler ropes in the interstices give some support to the extruded lead sheath. However, a common lead sheath over three cable cores plus some filler ropes is not a very stable design. These cables are sensitive to fatigue, and external forces or movements can damage the lead sheath. Also, these cables can be manufactured only in a short length and require many joints for long installation lengths. The 150 kV Java-Madura 3C oil-filled cable (commissioned 1987) has a common lead sheath and was produced in short drum lengths. These were transported to a jointing facility at the port where the cable pieces were jointed and loaded on-board the vessel [42].

3C-cables with individual lead sheath (denominated "S.L." for "single lead") have been used since many decades ago as submarine cables. Especially 3C cables with extruded insulation are almost always made with individual lead sheaths.

The individual lead concept has some important advantages over the common-lead concept:

- larger flexibility in the factory to joint individually sheathed cable cores
- better stability of lead sheaths because they are smaller in diameter and have a circular core underneath
- better stability during installation.

The advantage of a 3C cable with common lead sheath is the large oil channel provided by interstices in-between the individual cores. But for all practical applications these systems can be replaced by XLPE systems today without the need of oil channels or oil feeding.

Oil-filled submarine cables are still being produced for the EHV level – 500 kV and higher. Cables at this voltage levels are so massive that they cannot be produced as 3C cables.

As 3C oil-filled submarine cables are not produced anymore, this subject is not further elaborated here.

## *2.6.1 Choice Between One Three-Core and Three Single-Core Cables*

In some cases, the customer has the choice between one 3C cable and three 1C cables for the selected route. The best choice is not always obvious but depends on a large number of factors including cable route characteristics, installation and protection methods, factory facilities etc. The following considerations apply only to a.c. cables.

The armoring losses in 3C cables are significantly lower than in 1C cables because the magnetic fields from the three conductor currents cancel each other to a large extent. Therefore, 3C cables can be equipped with armoring of mild steel wires where 1C cables require more sophisticated armoring solutions. The lower losses in a 3C cable may allow for a smaller conductor to achieve the required ampacity. On the other hand, the heat transfer from a 3C cable is inferior to that from a group of spaced 1C cables. A detailed thermal calculation of both concepts will reveal the required conductor size for the 1C and 3C solution.

A 3C cable is in principle more expensive than the sum of three 1C cables on account of the additional factory step of lay-up of the three cable cores. The largest lay-up machines today can handle continuous cable core lengths of a few kilometres. For longer 3C cables, the lay-up machines need to be reloaded with new cable core making flexible factory joints necessary and thus adding to the costs.

Whereas 20 years ago 3C cables could be delivered only in short lengths, making many joints necessary, the situation is much different today. Depending on the equipment of the supplier, 3C cables for 100–170 kV can be delivered in continuous lengths of 50 km and beyond. A 50 km 115 kV 3C XLPE cable was supplied and installed in one piece in Saudi Arabia in 2004. The limitation today is set by the capacity of factory turntables and cable laying vessel (approx. 3000–7000 t).

The choice of 3C vs. 1C cables has a strong influence on the cable installation. A 3C cable can be installed in a single run, provided the vessel could load sufficient cable for the entire cable route. A group of 1C cables would require multiple installation runs including foregoing route survey and route clearing. 1C cables can often be installed and buried with lighter equipment than 3C cables because the cable's unit weight, bending stiffness, and bending radius are smaller.

It has been suggested to use 1C cables and install them simultaneously bundled in the same trench. Here, the drawbacks of both concepts (large armoring losses in 1C cables, bad cooling of 3C cables) are combined. Furthermore, cable laying vessels equipped for the simultaneous laying of three power cables are not easily available.

Another important aspect is the availability of the cable systems. It is common practice to install a group of 1C cables with some distance (up to several hundred metres) to avoid multiple damages. Also, it is common practice to install four cables for a three-phase circuit, or seven cables for two circuits, in order to have one spare. In case of a cable fault, the circuit can be restored within hours using the spare cable. The cable fault can then be repaired whenever practically possible. Loss of revenue can be avoided to a large extend. To achieve a similar redundancy for a 3C cable only a second 3C cable would do at considerable costs.

The question of single-core cable vs. three-core cables is further discussed in reference [43].

## 2.7 Two-Core Cables

Two-core cables have been used for HVDC systems where two conductors naturally occur. A development of the Mollerhoj cable has two fully insulated cable cores in a common lead sheath and has been used for the Danish part of the KontiSkan link. The pressurised oil filling provides a cavity-free paper insulation and can be operated at 70°C conductor temperature. Both cable cores are operated at the same voltage, and the return current goes through the sea electrodes. The latest submarine cable of this kind was the Kontek link between Denmark and Germany with 600 MW d.c. power in 1996.

The NorNed HVDC cable between the Netherlands and Norway has two cable cores with individual lead sheaths inside a common armouring (cf. Fig. 2.10). The cable cores are operated with opposite voltages, and there is no need for a sea electrode. The advantage of two-core HVDC cables with opposed currents is the almost total cancelling of the external magnetic field. This effect, however, can be achieved easier by a tight bundling of two single-core cables during laying.

A novel use of two-core submarine power cables may arise from suggestions to install 6-phase a.c. cable systems to bring massive offshore wind farm power to shore [44]. According to these suggestions, two cable cores with opposite phase would be combined into a two-core cable with common armoring. Magnetic losses in the steel armoring would be minimised.

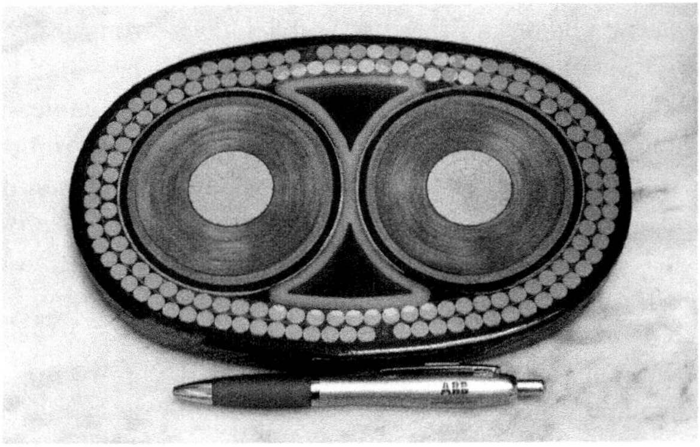


**Fig. 2.10** Two core 450 kV HVDC cable for the NorNed project

Flat two-core cables can be bent easily but in one direction. This property makes manufacturing, loading, and installing difficult and requires considerable investments in equipment.

## 2.8 Coaxial Cables

Coaxial cables have been used in earlier times to achieve concentric cables even in multiphase (two or three) systems (Fig. 2.11). Due to the difficulties of jointing and termination, they have been considered as impractical already more than 60 years ago [45]. Still, coaxial HVDC cables were installed in the Moyle Interconnector between Scotland and Ireland, in 2002.

**Fig. 2.11** Early two-conductor coaxial cable on display in Les Renardiers Research Center, Moret-sur-Loing, France

## 2.9 Optical Fibres Inside Submarine Power Cables

Optical fibres can be integrated into submarine power cables for several purposes:

1. Distributed measurement of temperature (DTS)
2. Data transmission
3. Measurement of cable strain or vibrations
4. Fault detection and location
5. Detection of changes in the cable route, e.g. changes in sediment cover over the cable.

Temperature measurement is based on short laser impulses launched into the fibre, and the back-scattered light (Raleigh scattering). The back-scattered light contains not only the original laser light wavelength but also faint side bands in the wavelength spectrum. The temperature can be determined from the ratio of the intensity of the spectral bands. As the back-scattered light from different distances in the fibre arrives at different times in the monitoring unit, a spatially resolved temperature measurement can be achieved. Different spectral phenomena can be used for temperature measurements. The temperature measurement unit provides a temperature profile along the length of the submarine cables. Some systems allow matching the profile into a map of the cable route for easier identification of hot-spots.

For a submarine power cable equipped with a DTS, a temperature profile of the cable route can be recorded some weeks after the cable commissioning, taking into account the initial warm-up. This reference temperature profile reflects the thermal characteristic along the cable route, such as soil cover, ambient temperature, etc. and can be used as a reference profile for comparison of subsequent measurements. The seafloor topography can change considerably due to tidal or other currents in the water, storms or human activities. Unwanted thick soil sediments above the cable, as well as unwanted de-burial of the cable result in changed thermal conditions, which can be detected by DTS when the cable is loaded. The data help the cable operator to decide on necessary corrective actions on the seafloor to restore cable protection or improve heat transfer from the cable.

Unfortunately, the DTS system suppliers use different terms to describe the accuracy of the temperature measurement systems. It is not always easy to compare the performance of different systems. Today, the largest length of temperature monitoring is about 30 km. Depending on the measurement system, the DTS needs one or two fibres, and there are different requirements on the need of connecting the far end of the fibres. If the far end is on shore, the loop can be made within a standard fibre joint box. However, if the monitored cable length ends somewhere *en route* in the sea, the connection of the far end can require the use of a joint box deep in the water. Sometimes very long lengths of DTS monitoring are stated, but this requires relay or amplifying boxes *en route*, which is rarely an option in submarine power cables.

Optical fibres can also be used for fault location. Standard OTDR (Optical Time Domain Reflectometry) methods use the attenuation of back-scattered laser impulses to determine the location of fibre damages. The measurement can easily be done from shore with standard instruments available at any fibre network installation company. Attenuation abnormalities and fibre breaks can be located with high accuracy (a few meters up to a few tens of meters). However, optical fibres can detect and locate power cable faults only to a certain extend. In a ruptured power cable, the fibre break will be clearly visible in the OTDR. But an anchor can damage the fibre (mostly laying in the cable armoring) without damaging the power cable core. A fibre break does not necessarily mean a broken power cable core, and a breakdown in the power cable insulation will not necessarily cause a detectable damage in the fibre.

The primary purpose of optical fibres, data communication, is also pursued in the combination with submarine power cables. Enormous submarine telecom cable networks embrace the planet. Extremely long hauls are covered, assisted by active or passive amplifier units. Though submarine power cables have been equipped with datacom fibres, a combination of submarine power cables and repeatered (=with amplifiers/repeaters) fibre-optical lines have not been tried yet. The reach of unrepeatered data transmission in increasing steadily. At standard data transmission rates of 5 or 10 GB/s for each fibre, repeater-less links can cover 200 km and more.

In-field cables in OWP's are made as three-core medium-voltage cables with integrated optical cables for monitoring and control of the turbines. The fibre cable sits inside the interstices between the cable cores. The short length of these cables (typically 200–1000 m) allows for unspliced optical cables with high fibre counts. The fibre attenuation is often less important in these short lengths. Optical cables with welded steel or copper sheaths and a diameter of 10–20 mm are commercially available with high fibre counts (96 fibres or more), and can be integrated into the interstices between the cable cores. However, the data flow between offshore WTG could easily be handled by a handful of fibres. Large fibre counts require high efforts for testing and splicing. In case of failure, the complete cable would be replaced rather than jointing both fibre and power elements.

Longer three-core cables for a.c. for high-voltage, too, can be equipped with optical cables in the interstices between the cores.

In single-core submarine cables, the optical fibres can be accommodated in the armoring. The fibres are housed in a stainless steel tube of 0.9–4 mm diameter. A plastic sheath on the stainless steel tube has a diameter equal to or slightly under the diameter of the armoring wires so that the element can be placed instead of one of the armoring wires (cf. Fig. 2.10). The plastic sheath prevents galvanic corrosion between the stainless steel tube and the armoring wires, and gives protection against squeezing of the tiny tube.

In the tough environment of a cable armoring machine, the metallic tubes must be handled with great care. Undue strain and bending of the tubes result in increased attenuation or even break of the fibres long before the steel tube is being damaged. By its own Young modulus, the plastic sheath can increase the maximum allowed tension by up to 25% depending on the sheath thickness and material specification. PVC, PE, PP, PA can be used for the sheath. Tiny armoring wires around the centre steel tube, embedded into the plastic sheath, can provide additional tensional strength.

Optical fibres located in the armoring of rather long single-core submarine cables are at risk of damage by external force. Damages would require the jointing not only of the optical element but also of the power cable core. Fibre and power cables are very different in characteristics and repair conditions so it might be prudent to separate the systems and lay separate power and fibre cables (maybe in one single operation).

## 2.10 Five Generic Cable Types

The large number of different submarine power cable species and their varieties may be confusing to those not deeply involved in the subject. It is one of the main tasks of this book to create a better understanding of possibilities and constraints of submarine power cables. To do so, five generic cable types are presented here in tabular form.[3] Application and prominent characteristics are summarised.

**Table 2.6** Five generic submarine power cable types

| | 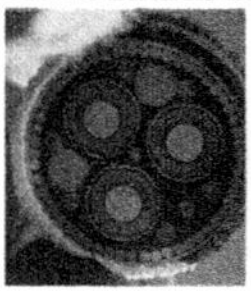 | 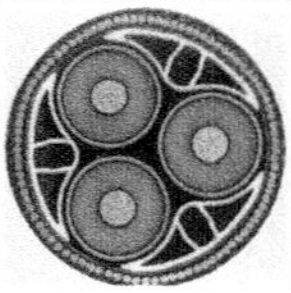 |  | 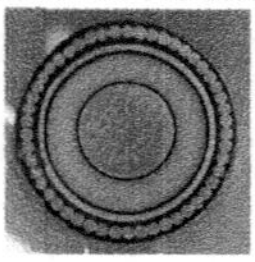 |  |
|---|---|---|---|---|---|
| Cable Type No. | 1 | 2 | 3 | 4 | 5 |
| Rated voltage $U_0$ | 33 kV a.c. | 150 kV a.c | 420 kV a.c. | 320 kV d.c. | 450 kV d.c. |
| Insulation | XLPE, EPR | XLPE | Oil/paper or XLPE | Extruded | Mass-impregnated |
| Typical application | Supply of small islands, connection of offshore WTG | Connection of islands with large population, OWP export cables | Crossing of rivers/straights with large transmission capacity | Long-distance connections of offshore platforms or wind parks | Long-distance connection of autonomous power grids |
| Max. length | 20–30 km | 70–150 km | < 50 km | >500 km | >500 km |
| Typical rating | 30 MW | 180 MW | 700 MW/ three cables | 1000 MW/ cable pair | 600 MW/ cable |

[3]The pictures in Table 2.6 are courtesy of NSW, Germany (cable no. 1), ABB, Sweden (cables no. 2, 4, 5) and Prysmian, Italy (cable no. 3).

The selected five cable types represent the majority of submarine power cables installed today. Other designs of course exist. Some of them are of historical interest, or have been used only in few cases. Others are highly specialised niche products such as power supply cables for subsea installations of the oil and gas industry.

Voltages and ratings are given for the five generic cables. That does not imply that these cables are standard cables. Submarine power cables are no off-the-shelf products. Most cable investments have different conditions that require variations of the cable design data. Multi-buyers, however, would gain from defining cable standards in order to reduce costs for redesign and type-testing.

## References


1. Ronström L et al. (2007). The Estlink HVDC Light® Transmission System. CIGRÉ Regional Meeting on Security and Reliability of Electric Power Systems, June 18–20, 2007, Tallinn, Estonia.
2. Hauge O et al. (1988). Performance of the ± 250 kV HVDC Skagerrak Submarine Cables, IEEE Transactions Power Delivery, Vol. 3, No. 1, January 1988, pp. 1–15.
3. Ingledow T et al. (1957). British Columbia–Vancouver Island 138 kV Submarine Power Cable. Proceedings of the Institution of Electrical Engineers, Vol. 104, Part A, No. 18, December 1957, London, UK.
4. Milliken H (1933). Electrical Cable. United States Patent Office, April 18, 1933, Serial-No. 1.904.162.
5. Milliken H (1940). Electrical Power Cable. United States Patent Office, January 16, 1940, Serial-No. 2.187.213.
6. Couderc D et al. (1996). Development and Testing of a 800-kV PPLP-Insulated Oil-Filled Cable and its Accessories. Cigré Session Paper 21/22-04, Cigré Session 1996, Paris, France.
7. Fromm U (2005). Optimized Conductors for XLPE Cables with a Large Cross-Section, European Transactions On Electrical Power, Vol. 15, pp. 109–121.
8. Bartnikas R (Ed.) (2000). Power and Communication Cables, IEEE Press Series on Power Engineering, ISBN 0-7803-1196-5.
9. Nicholls A and Steennis E (1990). Water Treeing, State of the Art. Cigré paper 15/21-02, at Cigré-session 1990, Paris, France.
10. Bernard G (1989). Working Group 09 of Study Committee 21: Application of Weibull Distribution to the Study of Power Cables Insulation. Electra No. 127, December 1989.
11. Jäger K-M et al. (2005). The Continuing Evolution of Semiconductive Materials for Power Cable Applications, IEEE Electric Insulation Magazine, Vol. 21, No. 1, pp. 20–34.
12. Brugg AG, Brug, Switzerland, High Voltage XLPE Systems, Technical User Guide. http://www.bruggcables.com/pdf/Brugg_Cables_User_Guide.pdf
13. Gorur R S et al. (2006). A Novel Approach for Prioritizing Maintenance of Underground Cables. PSERC Publication 06-40, Power Systems Engineering Research Center, Arizona State University, 577 Engineering Research Center, 2006 (http://www.pserc.org/cgi-pserc/getbig/publicatio/reports/2006report/gorur_cable_final_report_t23.pdf).
14. Brancato E L (1993). A Pathway to Multifactor Aging, IEEE Transactions on Electrical Insulation, Vol. 28, No. 5, pp. 820–825, October 1993.
15. Aras F et al. (2007). Aging of 154 kV Underground Power Cable Insulation, IEEE Electrical Insulation Magazine, Vol. 23, No. 5, September/October 2007, pp. 25–33.
16. Ross R et al. (1998). Application of Polyethylene Sheath and Swelling Powder Against Water Treeing. IEEE International Conferences on Conduction and Breakdown in Solid Dielectrics, June 22–25, 1998, Västerås, Sweden.

17. Reference website: www.crosssoundcable.com
18. Vahedy V (2006). Polymer Insulated High Voltage Cables, IEEE Electrical Insulation Magazine, Vol. 22, No. 3, May/June 2006, pp. 13–18.
19. Johnsen J et al. (1980). Discussion of Group 21, Cigré Session 1980. Submarine Power Cable Development in Norway, Cigré Session 1980, Paper 21–12.
20. McAllister D (1982). Electric Cables Handbook. Granada Technical Books Ltd, 1982, ISBN 0-246-11467-3.
21. US-Patent 2,531,917 filed 1946.
22. Matsunaga O et al. (2001). Leyte-Bohol Interconnection, 138 kV Oil-filled Submarine Cable Project in the Philippines. Hitachi Cable Review No. 20, August 2001.
23. Worzyk T (2007). 100 Years of High Voltage DC Links, Modern Power Systems, November 2007.
24. Worzyk T et al. (1997) Breakdown Voltage of Polypropylene Laminated Paper (PPLP) in Plain Samples and a Full Scale Cable. Conference on Electrical Insulation and Dielectric Phenomena, 1997. IEEE 1997 Annual Report.
25. Hata R (2006). Solid DC Submarine Cable Insulated with Polypropylene Laminated Paper (PPLP). SEI Technical Review, No. 62, June 2006.
26. Worzyk T (1996). Keine Gefahr bei Lecks in HGÜ-Seekabeln, Elektrizitätswirtschaft, Vol. 26.
27. Valenza D et al. (1995). HVDC Submarine Power Cable Systems-State of the Art and Future developments, Proceedings of 1995 International Conference on Energy Management and Power Delivery, Vol. 1, No. 21–23, pp. 283–287, November 1995.
28. Suen H (2007). Removal of 76 km 138 kV Gas-filled Submarine Cables with 50 Years Service. Presentation at IEEE PES Insulated Conductor Committee Meeting, Scottsdale, Arizona, USA, 2007.
29. Emerging Nanocomposite Dielectrics. Cigré Task Force Report TF D1.16.03, published in ELECTRA, 226, June 2006, pp. 24–32.
30. Green C et al. (2008). Nanodielectrics – How Much Do We Really Understand?, IEEE Electrical Insulation Magazine, Vol. 24, No. 4, July/August 2008, pp. 6–16.
31. Hunter P V (1956). Development of Power Cables. George Newnes Ltd, London.
32. Baur C (1903). Das elektrische Kabel. Springer Verlag, Berlin.
33. Foxall R G et al. (1984). Design, Manufacture and Installation of a 525 kV Alternating Current Submarine Cable Link from Mainland Canada to Vancouver Island. Paper 21-04, Cigré Session 1984, Paris, France.
34. Grosenbaugh M (2006). Development of Synthetic Fiber-Reinforced Electro-Optical-Mechanical Cables for Use with Moored Buoy Observatories, IEEE Journal of Oceanic Engineering, Vol. 31, No. 3, pp. 574–584.
35. Furugen M et al. (2002). Completion of Submarine Cable Lines Combining Low Environmental Impact with Low Cost, Furukawa Review, No. 21, pp. 44–49.
36. Paul W and Bentley D (2001). Conductor Survival in Lightweight Upper Ocean Working Cables. Proceedings of 11th International Offshore and Polar Engineering Conferences, Stavanger, Norway, pp. 708–714.
37. Li Y (2001). Corrosion Behaviour of Hot Dip Zinc and Zinc–Aluminium Coatings on Steel in Seawater, Bulletin Materials Science, Vol. 24, No. 4, pp. 355–360.
38. Henriksson S (1980). Properties of Materials in Sea Water – a Survey of Metallic Materials. Swedish Corrosion Institute, Reg. No. 1980:4 1980-06-03.
39. Bascom E C et al. (1998). Construction Features and Environmental Factors Influencing Corrosion of a Self-Contained Fluid-Filled Submarine Cable Circuit in Long Island Sound, IEEE Transactions On Power Delivery, Vol. 13, No. 3, July 1998, pp. 677–682.
40. The CIGRE Paper (1986). Methods to Prevent Mechanical Damage to Submarine Cables, presented by Cigré Working Group 21 as Session Paper 21-12 at the 1986 Cigré Session, Paris, France.
41. Elgh L et al. (1982). Repair of 400 kV ac Submarine Cable. Paper 21-05, Cigré Session 1982.

42. Galloway S J et al. (1990). 150 kV Java-Madure Submarine Cable System Interconnection, Power Engineering Journal, January 1990, pp. 7–15.
43. Gilbertson O (2000). Electric Cables for Power and Signal Transmission. John Wiley & Sons, New York.
44. Brakelmann H et al. (2007). Transmission Technologies for Collective Offshore Windfarm Connections. Proceedings of European Wind Energy Conference, 7–10 May 2007, Milan, Italy.
45. Domenach L (1947). La Technique Des Cables Électriques Isolés. Proceedings of a Conference Arranged by Centre de Perfectionnement Technique et Administratif du Chemin de fer Metropolitain de Paris, 1947 (in French).

# Chapter 3
# Design

## Contents



## 3.1 Thermal Design

The aim of thermal design of submarine power cables is to devise a conductor size, which transports the required power without exceeding design temperature limits of the cable or the environment. The thermal design of power cables has been described in many textbooks [1–3] and industrial standards [4] and shall not be repeated here in all details. Instead, this book is supposed to convey an understanding for the


T. Worzyk, *Submarine Power Cables*, Power Systems,
DOI 10.1007/978-3-642-01270-9_3, © Springer-Verlag Berlin Heidelberg 2009

principles of thermal design. Items particular to submarine cables are also touched. Project planners and decision makers may learn about some aspects of the thermal design and especially the aspects related to submarine power cables. For detailed calculations of a particular case, the "cookbook recipes" can be found in the textbooks or obtained from submarine power cable manufacturers.

The basis of all thermal design is a thermal model of the cable. During operation, heat is generated in parts of the cable construction (heat sources), which is transported to the outside of the cable and disappears into the ambient (heat sink). On its way from the source to the sink the heat is traveling through various layers in the cable and in the ambient. The thermal model tries to describe the real situation by a system of equations. With a better knowledge on the cable design and the ambient conditions, a more precise model can be generated. The thermal model must be able to describe continuous and transient conditions.

The cable operator wants to use his assets to achieve maximum benefit without jeopardizing the reliability or useful life of the asset. The benefit from a power cable lies in the accumulated transmitted energy and in the maximum power transmitted at any time. This must be achieved without exceeding the maximum operational limits of the cable.

The cable operator expects answers from the thermal model to at least the following questions:

- Which conductor cross section is necessary for the required transmission power under the given ambient conditions?
- Which short-term overload can I transmit, if the cable already has reached at a certain temperature?
- How does the maximum load change during the annual temperature variations?
- Which regular load pattern can be used under given thermal conditions?
- What happens if other cable operators install cables nearby?

Thermal models with various complexity are available for different purposes. Some simple cases, the dimensioning of a single-core HVDC cable and a pair of HVDC cables in a homogeneous seafloor, are described here to illustrate the basic idea. The symbols are corresponding to those used in the well-known industrial standard IEC 60287 [4].

### 3.1.1 Single-Core HVDC Cables

The ohmic losses in the conductor are the only heat source in an HVDC cable.[1] The loss $P_L$ can be described as

[1] Most HVDC links have a harmonic spectrum of overtones laying on top of the d.c. current. These overtones create some extra losses which however most often are very small, and are neglected here.

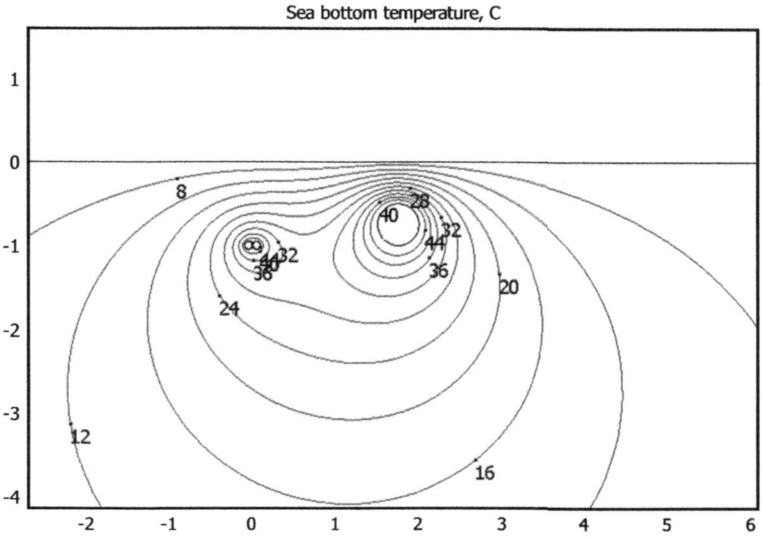


**Fig. 3.1** Equithermal lines in an installation of a pair of buried submarine cables (*left*) close to a pipeline (*center*) with 50°C surface temperature. The seafloor temperature at level 0 is assumed to be 4°C

$$P_L = I_C^2 \cdot R' \tag{3.1}$$

where the conductor resistance $R'$ is depending on the temperature:

$$R' = R_{20°C} \cdot (1 + \alpha(\Theta_C - 20°C)) \tag{3.2}$$

Both ohmic losses $P_L$ (W/m) and conductor resistance $R'$ (Ω/m) are given as per-unit-length of the cable. The conductor resistance at 20°C for various conductor sizes is given in Table 3.1. The values apply to conductors made from compressed round wires (IEC 60228 class 2).

Other conductor designs may have, with identical cross section, different resistance values due to other lay length, the presence of water-blocking compound, etc. HVDC cables with Conform-produced profiled conductor wires have often lower resistance values than those in Table 3.1. However, the values above may be used for the thermal design of HVDC cables as they constitute a conservative approximation.

Under steady-state conditions, all the heat generated in the conductor must flow to the outside of the cable. First the heat flows through the cylindrical dielectric insulation. The equation

$$\Delta\Theta = P_L \cdot T_1 \tag{3.3}$$

describes the relation between the temperature drop $\Delta\Theta$ across the insulation wall and the thermal resistance $T_1$ of the cylindrical insulation wall. For $T_1$:

**Table 3.1** Resistance values for copper and aluminium conductors according to IEC 60228, class 2

| | Copper | Aluminium |
|---|---|---|
| Specific electric resistivity@20°C, $\Omega$mm$^2$/m | 0.01786 | 0.02874 |
| Thermal coefficient of the specific electric resistivity @20°C, 1/K | 0.00392 | 0.0042 |
| Conductor resistance according to IEC 60228, Class 2. 20°C | $\Omega$/km | $\Omega$/km |
| 240 mm$^2$ | 0.0754 | 0.125 |
| 400 mm$^2$ | 0.0470 | 0.0778 |
| 500 mm$^2$ | 0.0366 | 0.0605 |
| 630 mm$^2$ | 0.0283 | 0.0469 |
| 800 mm$^2$ | 0.0221 | 0.0367 |
| 1000 mm$^2$ | 0.0176 | 0.0291 |
| 1200 mm$^2$ | 0.0151 | 0.0247 |
| 1600 mm$^2$ | 0.0113 | 0.0186 |
| 2000 mm$^2$ | 0.0090 | 0.0149 |

$$T_1 = \rho_T/(2\pi) \cdot \ln(D_o/D_i) \tag{3.4}$$

where $D_\text{o}$ is the outer diameter of the insulation screen and $D_i$ the diameter of the conductor. Although the semicon layers of the screens and the insulation have completely different electric properties they have similar thermal properties and therefore they are combined into $T_1$. The symbol $\rho_T$ denotes the thermal specific resistivity of the insulation material. Table 3.2 lists the specific thermal resistivity for some materials.

In a similar way, the thermal resistance of other cylindrical layers in the cable can be determined. For an arbitrary layer $n$:

$$\Delta\Theta_n = P_L \cdot T_{(n)} = \rho_T/(2\pi) \cdot \ln(D_o/D_i) \tag{3.5}$$

**Table 3.2** Thermal resistivity of cable design materials according to IEC 60287

| Insulation | Specific thermal resistivity K·m/W |
|---|---|
| Paper insulation, oil-filled cables | 5.0 |
| Paper insulation, mass impregnated | 6.0 (according to IEC 60287) |
| PPL | 5.5 |
| Polyethylene, XLPE or PE thermoplastic | 3.5 |
| Polyethylene for extruded HVDC cables | 3.5 |
| EPR for cables above 3 kV | 6.0 |
| Outer serving | |
| PE | 3.5 |
| PVC | 6.0 |
| Polypropylene yarn immersed in water | 3.7 (estimated) |

with $\Delta\Theta_n$ being the temperature drop over the $n$th layer. $T_{(n)}$ is the thermal resistance of the $n$th layer, $D_o$ the outer diameter and $D_i$ the inner diameter of the layer. In our cylindrical single-core cable the total temperature difference between the conductor and the cable surface now is

$$\Delta\Theta = \sum_n (P_L \cdot T_{(n)}) = P_L \sum_n T_{(n)} = I_C^2 R' \sum_n T_{(n)} \tag{3.6}$$

where $I_C$ is the conductor current. The sum runs over all layers in the cable. The thermal model of IEC 60287 defines four thermal layers $T_1$ to $T_4$ for the cable and its ambient soil as follows:

$T_1$: Thermal resistance of the dielectric insulation as defined in Eq. 3.4.
$T_2$: Thermal resistance between metallic screen/sheath and armoring. This layer also includes bedding layers under the armoring.

$$T_2 = \rho_T/(2\pi) \cdot In(1 + 2\,t_2/D_s). \tag{3.7}$$

where $t_2$ is the layer thickness and $D_s$ is the outer diameter of the metallic screen/sheath. The extruded plastic sheath over the lead sheath of submarine cables is accounted for in $T_2$. The bedding is often bituminized fabric tape, sometimes arranged with overlap so that their thermal behavior is difficult to describe. Fortunately the bedding is less than 1 mm in thickness and contributes so little to the total thermal resistance that possible errors in the assumed thermal resistivity are acceptable.

$T_3$: Thermal resistance of the outer sheath (serving) over the armoring.

$$T_3 = T/(2\pi) \cdot In(1 + 2\,t_3/D'_a). \tag{3.8}$$

where $D'_a$ is the outer diameter of the armoring and $t_3$ is the layer thickness of the serving/outer sheath. The $\rho_T$ values for some serving materials can be found in Table 3.2.

Metallic screens covering the complete cable can be neglected, as their thermal resistivity is negligible in comparison with that of other cable construction materials. In this context we assume: $\rho_T$ (metal) = 0, $T_{(n)}$ (metal) = 0.

Flat wire armoring with dense laying acts as a consolidated metallic layer and has virtually no thermal resistance compared to other cable layers. An armoring made from round wires, however, is a mixture of metal with low thermal resistivity and air/water/bitumen having a higher thermal resistivity. A detailed differential analysis shows that the composed thermal resistance of a layer of densely packed round wires is at the most 1.5 times the thermal resistance of a solid metal layer of the same thickness. Given the low thermal resistance of metals in the cable construction, we can neglect the thermal resistance of densely packed wire armoring no matter if the interstices between the wires are filled with air or water.

Now, the thermal resistances of all cable layers have been computed. For buried submarine power cables the heat flow continues through the seafloor soil. The accumulated thermal resistance between the buried cable and the sea floor is designated $T_4$ according to the symbols used in IEC 60827.

#### 3.1.1.1 Single Buried Cable

For a single cable buried in homogeneous sea bottom soil at depth $L$ the thermal resistance $T_4$ between the cable surface and the seafloor is:

$$T_4 = \rho_T/(2\pi) \cdot In(2\,u) \tag{3.9}$$

where $u = 2L/D_e$, relating the burial depth $L$ to the cable diameter $D_e$. The burial depth $L$ is defined as the vertical distance between the seafloor and the cable axis (cf. Fig. 3.2). Equation 3.9 is a good approximation for $u > 10$ and applies for most buried cables. For a shallower burial depth of the cable, IEC 60827 suggests more complicated formulae. Typical values for the thermal resistivity $\rho_T$ of the seafloor soil are given in Table 3.6 in Sect. 3.1.3.1.

#### 3.1.1.2 A Pair of Buried Cables

For two identical equally-loaded HVDC cables laid in the same depth, $T_4$ can be calculated as:

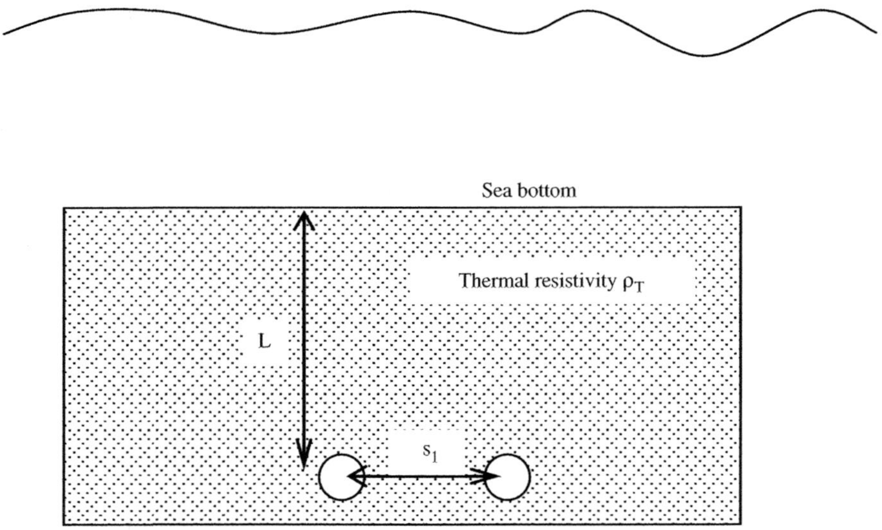


**Fig. 3.2** Symbols used for the calculation of $T_4$

$$T_4 = \frac{\rho_T}{2\pi}\left(\ln\left(u+\sqrt{u^2-1}\right)+\frac{1}{2}\ln\left(1+\left(\frac{2L}{s_1}\right)^2\right)\right) \tag{3.10}$$

For $u > 10$ (which is most often true in submarine installations) the term $(u + \sqrt{(u^2-1)})$ can be replaced with $(2u)$.

$D_e$ outer Diameter of the cable/s
$L$ laying depth, as measured from the sea floor surface to the cable axis
$s_1$ axis distance of the cable pair
$\rho_T$ specific thermal resistivity of the soil material.

The important $T_4$ value for a pair of identical equally loaded cables can also be found from Fig. 3.3. The figure shows the ratio of $T_4$ normalized for $\rho_T$ as a function of $L/D_e$, i.e. the relation between burial depth and the cable diameter. The parameter is the ratio between the burial depth and the cable spacing $s_1$. Knowing the laying parameters and the cable diameter, the value $T_4/\rho_T$ can be found from the graph. Finally $T_4$ is obtained by multiplying with $\rho_T$. Most submarine HVDC cable links can be assessed with Eq. 3.10 as both cables of a HVDC pair are equally loaded and buried at the same depth.[2] Local moderate differences in the burial depth have only little influence on $T_4$.

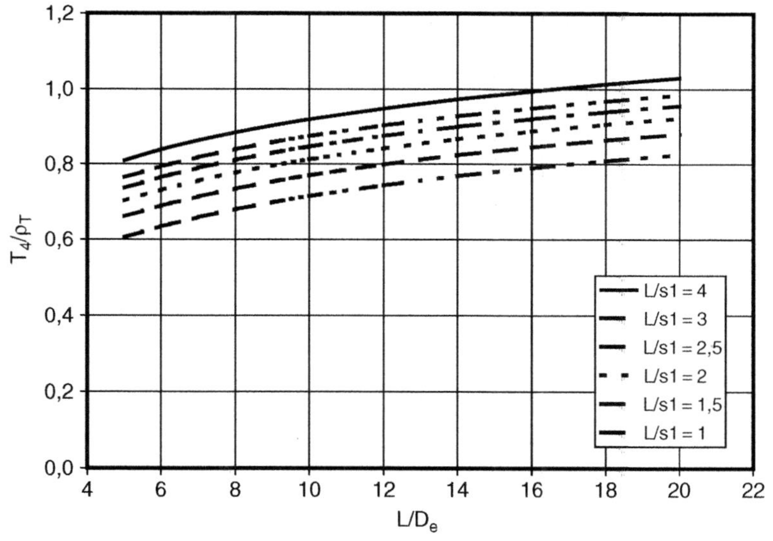


**Fig. 3.3** Thermal resistance $T_4$ of the soil (related to $\rho_T$) vs. the laying geometry. Denotations are explained in the text

[2]Monopolar HVDC systems with one pole cable and one metallic return cable should be calculated with different equations, as the cables are not identical.

Now we have calculated the thermal resistances $T_1$ to $T_4$ and can continue to calculate the ampacity[3] of the HVDC cable as follows:

$$I = \left[\frac{\Delta\Theta}{R'\left(T_1 + T_2 + T_3 + T_4\right)}\right]^{0.5} \tag{3.11}$$

Here $\Delta\Theta$ is the maximum allowable temperature difference between the conductor temperature $\Theta_c$ and the undisturbed ambient soil temperature $\Theta_{\text{amb}}$:

$$\Delta\Theta = \Theta_c - \Theta_{\text{amb}}$$

As an example the ampacity of a pair of extruded HVDC cables (Cable type 4) is determined (Table 3.3 and Table 3.4):

The thermal design of most submarine HVDC cables and cable pairs can be performed according to the simple model of Eqs. 3.9 and 3.10. A more detailed calculation is required when both conductors of a HVDC link are combined into one cable, such as in the Kontek, NorNed and Moyle Interconnector.

**Table 3.3** Thermal properties of a pair of extruded HVDC cables

| Cable design data | |
|---|---|
| Conductor 1200 mm$^2$ Cu | $D = 39.9$ mm |
| Conductor resistance at 70°C | $R(70°\text{C}) = 0.0182$ Ω/km |
| Maximum allowed conductor temperature | 70°C |
| Insulation system (1.2 mm semicon, 12 mm insulation, 1.2 mm semicon) | $T_1 = 0.328$ K·m/W |
| Lead sheath, thickness 2.5 mm | |
| Plastic sheath, thickness 2.5 mm | $T_2 = 0.036$ K·m/W |
| Armoring round wires, 5 mm wire dia | |
| Outer serving yarn thickness 4.0 mm | $T_3 = 0.087$ K·m/W |

Conductor tapes, bedding tapes etc are not mentioned but included in the calculations.

**Table 3.4** Thermal rating of a pair of extruded HVDC cables

| Cable installation data | |
|---|---|
| Laying depth under the sea floor | 1.5 m |
| Spacing between the cables | 5 m |
| Thermal resistivity of the sea floor | 0.8 K·m/W |
| $T_4$ | 0.5423 K·m/W |
| Ambient temperature in the sea floor | 8°C |
| Current rating | 1854 A per cable |

[3]The word ampacity denotes the current carrying capability of a cable and has been coined by W. A. DelMar of Phelps Dodge Wire&Cables in 1951 [2].

Mainly two strategies are used to increase the ampacity of the cable system to meet the project requirements: changing the laying parameters (depth and spacing), and changing of the conductor resistance. The use of a larger conductor in order to reduce the conductor losses will cause minor changes in the $T_1$ to $T_4$ values. Therefore, the calculation of the optimum cable system is an iterative process.

### *3.1.2 a.c. Cables*

The calculation of a.c. ampacities is much more complex compared to d.c. cables. Additional losses are created in the conductor and the armoring as a result of the alternating current.

#### 3.1.2.1 Conductor Losses

The magnetic alternating field around the conductor current causes the skin effect, by which the current density is low in the centre of the conductor, and high in the outer regions of the conductor. The useful conductor area is reduced, and the effective conductor resistance is increased. The skin effect is more pronounced with increasing conductor areas. Furthermore, the skin effect is depending on the resistivity of the conductor material, the conductor design, and the power frequency. These influences are summarized into the skin effect factor $y_s$. Since the skin effect adds to conductor losses, the cable ampacity is decreased. Skillful but expensive conductor design can reduce the skin effect considerably especially for large cross section conductors.

Another magnetic effect, the proximity effect, is caused by the proximity of the conductors in a three-phase system. Under the influence of a neighbor conductor, the current strives to concentrate in current paths as far away as possible from the disturbing neighbor. The current density in the conductor becomes inhomogeneous, rendering the conductor portions nearest to the neighbor less useful for current transport. This effect is mostly pronounced for closely spaced large ampacity conductors such as three-phase a.c. cables. The resulting apparent resistance is expressed:

$$R = R'(1 + y_s + y_p) \tag{3.12}$$

where

$R$ is the a.c. resistance of the conductor at maximum operating temperature
$R'$ is the d.c. resistance of the conductor at maximum operating temperature
$y_s$ is the skin effect factor
$y_p$ is the proximity effect factor.

The a.c. resistance $R$ (Ω/m) must be used for the calculation of the conductor losses. The factors $y_s$ and $y_p$ depend on the conductor material and design (stranding, lay-length, segmental or not), impregnated or not, and the power frequency.

Elaborate formulae to calculate $y_s$ and $y_p$ for various cable configurations are given in [2, 3], and shall not be repeated here. For the practical purpose of project planning, it is often enough to know that $(1 + y_s + y_p)$ for small cross section (500 mm$^2$) most often is below 1.1, i.e. skin and proximity effect increase the effective a.c. resistance of the conductor by less than 10%. Given the uncertainty of the thermal resistivity of the soil, it may be considered to neglect the skin and proximity contribution with small conductors. For very large conductor sizes ($\geq$2000 mm$^2$), however, skin and proximity effect may enlarge the a.c. resistance with more than 30%.

#### 3.1.2.2 Dielectric Losses

The cable insulation is a dielectric material and can be modeled as combination of a capacitance and a resistance in parallel between the conductor and the grounded screen. Applying a voltage to the conductor results in a capacitive and a resistive current. The resistive current is in phase with the voltage while the capacitive current is shifted with 90°. The resistive current is a loss current generating heat in the insulation. The ratio between resistive and capacitive current is called the loss angle tan δ:

$$\tan\delta = \frac{|I_r|}{|I_c|} = \frac{1}{R_i C\omega} \tag{3.13}$$

where $R_i$ is the resistance of a 1-m-piece of cable insulation (Ω.m) and $C$ is the capacitance per meter (F/m). $\omega = 2\pi f$ is the angular frequency of the a.c. voltage. The cable capacitance $C$ can be calculated as:

$$C = \frac{\varepsilon_0 \varepsilon_r}{18 \ln\left(\frac{D_i}{d_c}\right)} \tag{3.14}$$

where $D_i$ is the diameter of the insulation, $d_c$ the diameter of the conductor screen, and $\varepsilon_r$ is the relative dielectric constant of the insulation material (Table 3.5).

Now, the dielectric losses in the insulation can be calculated as:

$$W_d = \omega C U_0^2 \tan\delta \tag{3.15}$$

**Table 3.5** Dielectric properties of high-voltage cable insulation materials. tan δ is depending on the temperature. A more comprehensive list is in Chap. 12.

| IEC 60287 | Dielectric loss factor tan δ | Dielectric constant $\varepsilon_r$ |
|---|---|---|
| XLPE > 18/30 kV | 0.001 | 2.5 |
| EPR | 0.005–0.020 | 3 |
| Oil/paper > 87 kV | 0.0033 | 3.6 |
| Mass-impregnated | 0.01 | 4 |

The dielectric losses grow with the square of the cable voltage and are relevant only at higher voltage levels. According to IEC 60287, they can be neglected for XLPE cables under 127 kV. For oil-filled submarine power cables being used predominantly for high and extra high voltages, the dielectric losses must not be neglected because of the higher $\varepsilon$ and tan $\delta$ values. The dielectric losses are not depending on the arrangement of the conductors in single-core or three-core cables, nor on the design of the conductor.

### 3.1.2.3 Screen Losses

The alternating magnetic field around the conductor generates circulating and eddy currents in the metallic screen and armoring. These currents contribute to heat generation and reduce the cable ampacity. Four different types of induced current losses are considered according to the terminology of IEC 60287:

$\lambda'_1$ denotes the losses due to circulating currents in the screen/sheath.[4] Circulating screen/sheath currents occur when the screen/sheath is grounded on both sides of the cable. The losses are depending on the screen/sheath resistance, the screen/sheath inductance, and the arrangement of the phases in relation to each other. Circulating screen/sheath currents can achieve high amplitudes (up to the value of the conductor current) and add a heat source in the cable. The ampacity is reduced considerably. Especially single-core a.c. submarine cables suffer from screen/sheath losses. Loss-reducing measures such as cross-bonding or single-side bonding which are common in onshore cable installations can hardly be arranged for submarine cables.

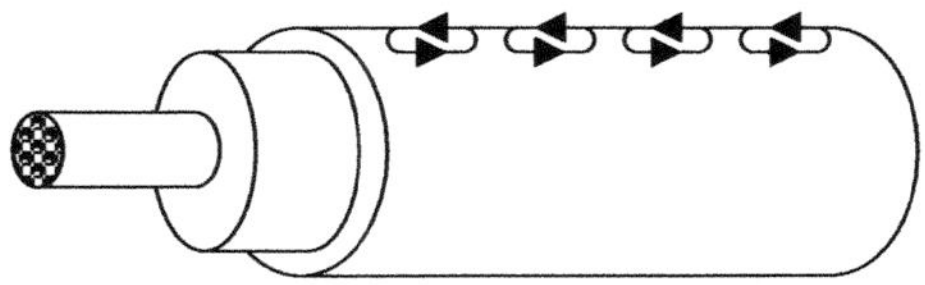

$\lambda''_1$ denotes the losses due to eddy currents in the screen/sheath. The alternatingconductor current generates an emf in the metallic screen or sheath. The emf drives eddy currents locally inside the metallic screen/sheath. The amplitude of the eddy currents is strongly dependant on the thickness of the screen/sheath. The specific resistivity of the materials involved and geometric factors play an important roll. In many cases $\lambda''_1$ can be neglected in comparison to the $\lambda'_1$ losses. However, $\lambda''_1$ should be taken into account for a.c. cables with large segmental conductors, and cables where the $\lambda'_1$ losses have largely been reduced by cross-bonding or single-side bonding.

$\lambda'_2$ denotes the losses due to circulating currents in the armoring and metallic protection pipes, if applicable.[5] $\lambda'_2$ is strongly depending on the total resistance of the armoring layer. Not only the specific resistivity of the armoring material but

[4]"Screen/sheath" includes metallic wire screens (most often Cu wires) and extruded or welded metallic sheathes, but not the armoring.

[5]Sometimes parts of the cable are protected with cast iron half-pipes or similar after installation. This should be taken into account for the cable design.

also the lay-length and contact points between individual wires influence the eddy current losses. Steel wire armorings are particularly affected by eddy current losses as the magnetic material attracts and concentrates the magnetic induction.

An a.c. cable with a low-resistance metallic screen/sheath has a rather high induced screen/sheath current opposite in direction to the conductor current. The magnetic fields from the conductor and the opposite screen current partly cancel each other. For this reason the magnetic effects on the armoring and external pipes are strongly reduced. Single-core a.c. submarine cables sometimes have a large copper screen able to carry the full screen current. As the screen current is opposite to the conductor current, the resulting magnetic field outside the screen is very low allowing for the use of a magnetic steel armoring without excessive armoring losses.

In three-phase a.c. cables the magnetic fields from the individual phases cancel each other to a large extend, which leads to low magnetic losses in the common armoring. Still, the armoring losses cannot be neglected.

$\lambda''_2$ denotes the losses due to eddy currents in the armoring and metallic protection pipes, if applicable.

All loss factors $\lambda'_1$, $\lambda''_1$, $\lambda'_2$ and $\lambda''_2$ are expressed as a factor relating the losses to the conductor losses. The calculation of the factors is explained for many different cable configurations in [2]. A concise tabular summary is found in [3], all based on the industrial standard IEC 60287, which is commonly accepted as the basis for the thermal design. However, the IEC formulae and factors are sometimes based on empirical values and can be subject to reviews.

For the common type of submarine a.c. power cables having three cores with individual lead sheath and common armoring, the loss factors $\lambda'_1$ and $\lambda''_1$ are as follows:

$$\lambda'_1 = \frac{R'_S}{R'} \cdot \frac{1.7}{1 + \left(\frac{R'_S}{X'}\right)^2} \tag{3.16}$$

where $X'$ is the reactance of the sheath:

$$X' = 2\omega \cdot 10^{-7} \cdot \ln\left(\frac{2s}{d}\right), \Omega/m \tag{3.17}$$

$R'_S$ is the resistance per meter of the lead sheath, and $R'$ is the resistance per meter of the conductor, $s$ is the axial distance of the conductors and $d$ the average sheath diameter. Unless we deal with large segment conductors (which rarely are used in three-core cables), we can assume

$$\lambda''_1 = 0. \tag{3.18}$$

For the same cable type ($\leq 400$ mm$^2$) with common steel wire armoring the losses in the armoring can be summarized as:

$$\lambda'_2 + \lambda''_2 = \lambda_2 = 1.23 \cdot \frac{R'_B}{R'} \cdot \left(\frac{2c}{d_B}\right)^2 \cdot \frac{1}{1 + \left(\frac{3.48R'_B}{\omega\mu_0}\right)^2} \cdot \left(1 - \frac{R'_S}{R'} \cdot \frac{1}{1 + \left(\frac{R'_S}{X'}\right)^2}\right) \tag{3.19}$$

where:

$c$ distance between cable axis and conductor axis
$d_B$ average diameter of the armoring layer
$R_B$ resistance per meter of the armoring
$R'_S$ resistance per meter of the lead sheath
$R'$ resistance per meter of the conductor
$s$ axial distance of the conductors
$d$ average sheath diameter.

#### 3.1.2.4 a.c. Cable Ampacity

Three-core a.c. cables have no coaxial geometry, which makes the calculation of the thermal resistivity between conductor and cable surface more complex. While the cable cores (conductor, insulation system, and screen/sheath) are coaxial and can be treated as in single core cables, thermal resistance between the cable core and the common armoring cannot be treated correctly with IEC 60287 methods. The materials in the interstices between the cable cores play an important roll for the heat transport to the ambient. Assemblies of circular polymeric filler ropes, extruded plastic profiles with hollow cores, or even lead profiles are put into the interstices, having much different thermal properties. Extruded plastic profiles with hollow cores are air-filled during manufacturing, but they might fill with water after installation. In these cases, the heat flow and temperature field inside and outside the cable can easily be calculated with commercial FEM software. The mesh generating routines in most FEM software is adaptive with sufficiently high dynamic so that both the temperatures in the small cable details and in the wide cable surrounding can be calculated.

Now the elements necessary to calculate the ampacity of a.c. cables have been collected. The ampacity is expressed as:

$$i = \left[\frac{\Delta\Theta - W_d\,[0.5\,T_1 + n\,(T_2 + T_3 + T_4)]}{RT_1 + nR\,(1 + \lambda_1) + nR\,(1 + \lambda_1 + \lambda_2)\,(T_3 + T_4)}\right]^{0.5} \tag{3.20}$$

where $n$ is the number of conductors in the cables (one or three). The a.c. resistance $R$ is calculated according to Eq. 3.12. In three-core cables the proximity effect is more accentuated and the a.c. resistance is affected accordingly. The thermal resistances $T_1$ and $T_3$ can be calculated according to Eqs. 3.4 and 3.8. The calculation of $T_2$ for single-core cables can be done according to Eq. 3.7, while the same value for three-core cables is depending on the design of the metallic sheath.

Again, conductor losses and the thermal ambient of the cable are the most dominant factors in the thermal design. Additionally, for a.c. cables the control of losses in screen/sheath and armoring is important to achieve a high ampacity, especially at higher transmission power.

## 3.1.3 Other Factors for the Thermal Design

### 3.1.3.1 Transient Conditions

The calculation models shown so far are used to calculate ampacities for steady-state conditions reflecting the case of continuous constant load over long time. In real life, cables are rarely operated at constant load over a longer time. Knowing that, the thermal design of a given submarine cable project can be reconsidered. For this, it is important to know how fast the cable temperature increases after the load is switched on, and after which time the cable reaches a steady-state condition under the impact of a constant load. Taking into consideration the load variations over time, one can perform a transient cable rating.

The mathematical challenge is to express the temperature response of the different cable layers and its surrounding on a step function representing the power switch-on. In the 1950s, when computers were not commonly available, many researchers suggested analytical solutions before Neher/McGrath published their famous papers in 1964 [5, 6], which subsequently influenced the IEC standard 60287.

The thermal models of power cables represent the various layers of the cable by thermal resistances and thermal capacitances. The heat dissipated in the conductor travels through the thermal resistance of the insulation and the outer layers towards the ambient soil.[6] The analogous network shown in Fig. 3.4 is called a Cauer-type RC ladder network. Cauer networks are often used to calculate heat flow phenomena in semiconductor devices.

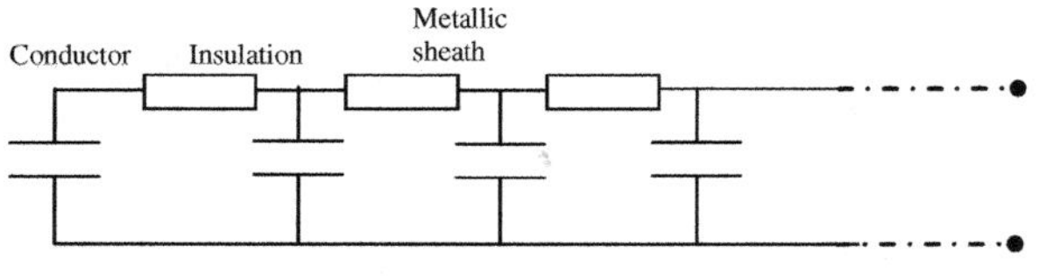


**Fig. 3.4** Cauer-type ladder network representing a cable construction

The ladder network consists of the thermal resistances $R$ of the various cable layers and the ambient soil, and capacitances $C$, which represent the thermal capacities of the different cable layers and the ambient soil. The thermal capacitance of a given volume $V$ of a material with the specific heat $c_p$ is:

[6] In the electric analogy the conductor loss corresponds to a current source, the heat flow from the conductor outwards corresponds to an electric current, the thermal resistance of each layer corresponds to an electric resistance, and the thermal capacitance of each layer corresponds to an electric capacitance.

$$C = c_p \cdot V \cdot \rho_G$$

where $\rho_G$ is the density of the material.

Let us consider a simplified fictitious cable, only consisting of a 1200 mm$^2$ copper conductor and a 12 mm polymeric insulation, directly submersed into water. The outer surface of the cable is kept at 10°C by ambient water. At $t = 0$ the load is switched on, generating a constant conductor loss of 30 W/m irrespectively the conductor temperature. The conductor losses heat up the thermal capacitance representing the conductor (solid line in Fig. 3.5), and the conductor temperature follows a logarithmic curve until a steady-state condition at about 18°C. The increasing temperature difference between the conductor and the ambient causes an ever-increasing heat flux (dotted line in Fig. 3.5) through the thermal resistance representing the insulation. The increasing heat loss through the insulation reduces the thermal power available for heating up the conductor, and the steepness of the temperature rise curve declines. A classical thermal time constant $\tau = R.C$ can be defined for this simple system. The time constant of this system is 1220 s as indicated by an arrow. After 1220 s, the temperature rise has reached 63% of its final value.

In a real cable installation, there are more thermal resistances and thermal capacitances coupled into a ladder network. Some layers, such as a lead sheath, can simply be represented by a thermal capacitance, while their thermal resistance can be neglected due to the high thermal conductivity of the metal. Insulation layers of a few millimetres thickness can be modelled by a single thermal resistance and a single thermal capacitance. Thicker insulation layers must be subdivided in a number of "onion shells" each of them with its own resistance and capacitance. Many computer-based models of the thermal ladder comprise 20 layers or more. A further complication is that in many high-voltage a.c. cables heat sources exist outside

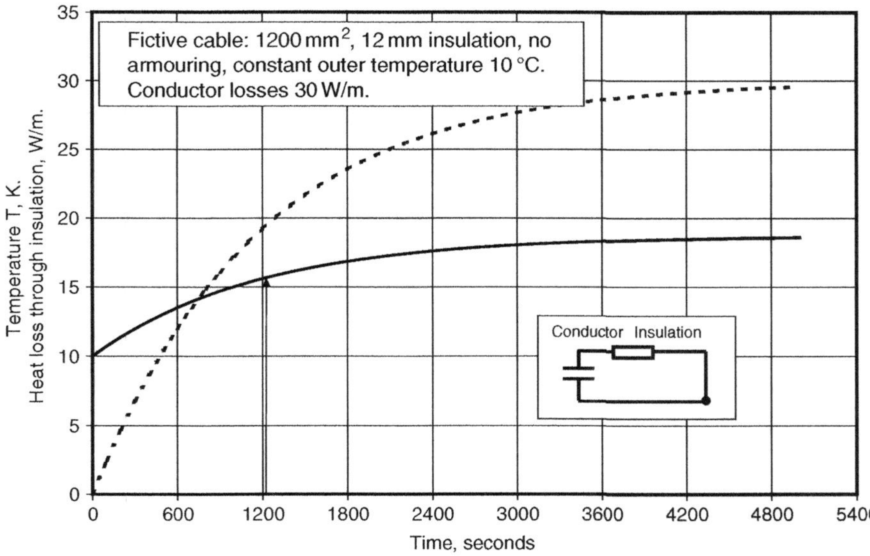


**Fig. 3.5** Temperature evolution (*solid line*) and heat flow through the insulation (*dotted line*) in a fictive simplified single-core cable

the conductor, such as dielectric losses and sheath/screen/armoring losses. It is not possible to define a classical thermal time constant for these complicated systems, as the step response is not longer a simple logarithmic rise such as in Fig. 3.5.

The ambitious reader may find exhaustive reading on the subject in [2, 7]. However, a few further considerations are important for the understanding of the cable behaviour and rating.

#### 3.1.3.2 Temporary Overload

Today, submarine power cables are usually buried 1–3 m down in the seafloor. The seafloor temperature will rise slowly, heated by the cable losses. The constant ambient temperature assumption in the simple model of Fig. 3.5 is not valid anymore. Since the seafloor has a large thermal capacitance and a large thermal resistance due to its sheer size, it can take weeks or months to warm up the seafloor around the cable to steady-state conditions. Although a time constant in a classical sense can no longer be defined, it can be useful to use two "quasi"-time constants to characterize the complex system of cable and surrounding soil. The thermal time constant $\tau_C$ of the cable describes the response of the conductor temperature to a step-formed load change, while the soil time constant $\tau_S$ describes the response of the soil temperature to a long-term load in the cable. Even for the most massive submarine cables, $\tau_C$ is in the order of 0.5–2 h. In contrary, $\tau_S$ is in the order of weeks or months. This dual time response behavior has some important consequences.

The evolution of temperature in and around a submarine cable is illustrated in Fig. 3.6. It starts with the "all-cold" situation (curve No. 1) where the cable and

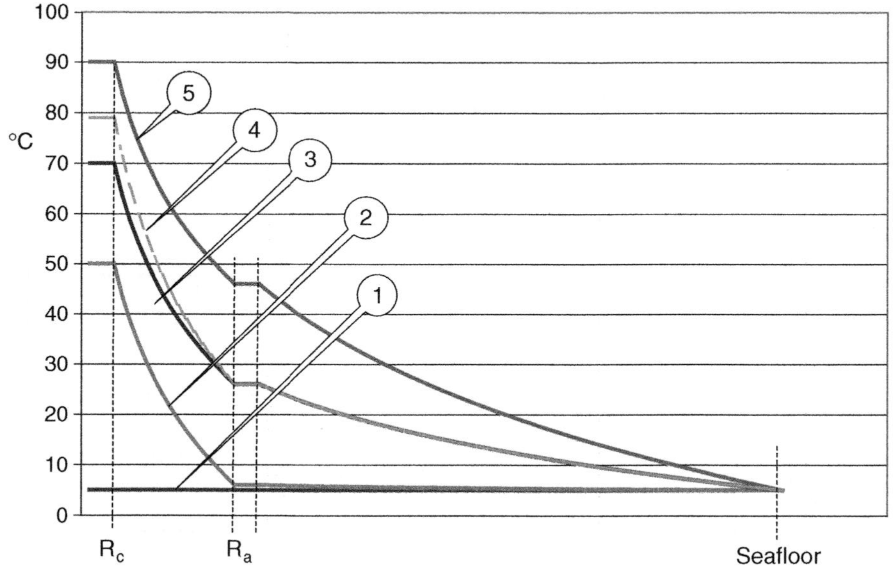


**Fig. 3.6** Temperature profile in a single-core cable and surrounding soil

the seafloor have the same undisturbed temperature, being 5°C in our example. The x-axis of Fig. 3.6 indicates the distance from the conductor, where $R_c$ denotes the conductor radius, $R_a$ the radius of the armoring, and the label "Seafloor" indicates the position of the surface of the sea bottom.

Now we put full rated load on the cable from "all-cold" condition, and the cable will reach a first steady-state after 3–4 times $\tau_c$, i.e. a few hours. This situation is indicated in curve 2 of Fig. 3.6. It is also visible that the temperature curve inside the conductor is flat because the metal equalizes all internal temperature gradients. The conductor temperature reaches 50°C in this example. During this time, the ambient soil (between $R_a$ and "Seafloor") changes its temperature only very little.

Under continued rated load, the soil temperature rises slowly as a result of the dissipated heat from the cable. At the same pace as the soil temperature next to the cable surface rises, the cable conductor follows suit. An intermediate situation is depicted as curve no. 3 in Fig. 3.6. As the soil and thus the cable armoring (at $R_a$) slowly increase in temperature, the conductor temperature keeps pace by always holding 45 K over the cable armoring. The temperature difference between armoring (at $R_a$) and the conductor stays constant, as this is only dictated by the conductor heat losses, which are assumed to be constant.[7] Finally, still under constant rated load, the system arrives at a steady-state indicated by curve no. 5 in Fig. 3.6. In the steady-state value the conductor temperature has arrived at 90°C and the cable surface (armoring) temperature is at 46°C.[8]

Let us go back to the situation of curve no. 3 in Fig. 3.6. The cable had been operated at full nominal load for days or even some weeks. Temperatures have not reached their final steady-state value yet. It is obvious from the graph that the conductor temperature is below the maximum temperature that it is designed for (90°C), and that it would reach only some weeks later. Meanwhile, we could use the thermal reserve to run temporary overload on the cable. Higher current means larger ohmic losses in the conductor, a larger heat flow through the insulation, and a larger temperature drop over the insulation. This renders a steeper temperature curve between $R_c$ and $R_a$ as indicated in curve no. 4 in Fig. 3.6. The overload can only be run 0.5–2 h until the conductor temperature reaches the design limit and must be reduced subsequently. Different restrictions may apply for HVDC cables as they usually have a maximum allowed temperature difference across the insulation.

A different starting case for a possible temporary overload occurs when the cable has been run with reduced load for a long time. In the following section, the possible overload for short periods is estimated with some easy equations. Recall the calculation of the maximum continuous current $I_{\text{cont}}$ according to Eq. 3.21:

$$I_{\text{cont}} = \left[\frac{\Delta\Theta}{R'\,(T_1 + T_2 + T_3 + T_4)}\right]^{0.5} \tag{3.21}$$

---

[7] For the sake of simplicity, we neglect here that the conductor resistance and hence the losses increase with the conductor temperature.

[8] This value is only valid for this example. Cable surface temperatures are depending strongly on the specific case.

The steady-state conductor temperature $\Theta_c$ is according to Eq. 3.21:

$$\Theta_c = \Theta_{\text{amb}} + I^2_{\text{cont}} \cdot R' \cdot \sum \cdot T_i \tag{3.22}$$

When the same cable had been running at 75% load ($I_{75\%} = 0.75\ I_{\text{cont}}$) the steady-state temperature, accordingly, would have been:

$$\Theta_{75\%} = \Theta_{\text{amb}} + I^2_{75\%} \cdot R' \cdot \sum \cdot T_i = \Theta_{\text{amb}} + 0.75^2 I^2_{\text{cont}} \cdot R' \cdot \sum \cdot T_i \tag{3.23}$$

The steady-state temperature $\Theta_{75\%}$ at 75% load is lower than the full load temperature and the difference is $\Theta_c - \Theta_{75\%}$. Now, we switch on the overload current $I_{\text{overload}}$. For a short-term overload (say 10 min), we may assume a quasi-adiabatic situation where the excess heat is only used to heat the conductor but the heat flow through the insulation is maintained at the level of the 75% constant load prior to overload. For an overload time of 10 min = 600 s, we can establish:

$$(I^2_{overload} \cdot R' - I^2_{75\%} \cdot R') \cdot 600\, s = c_p \cdot m_{Cu} \cdot (\Theta_c - \Theta_{75\%}) \tag{3.24}$$

The left side of the equation is the difference between the loss power $I^2{}_{\text{overload}}\%R'$ and the power outflow $I^2{}_{75\%}{}'R'$ through the insulation, multiplied with the duration of the overload current. This is the excess energy that is consumed by the heating of the conductor. $c_p$=393 J/(kg K) is the specific heat of copper, and $m_{Cu}$ is the mass of one meter of the copper conductor. For a cable with a 1200 $\text{mm}^2$ Cu conductor we find that $I_{\text{overload}}$ (10 min) = 3513 A. This is indeed an amazing figure of overload after a long continuous load of 75% of nominal load.

The temporary overload calculation according to Eq. 3.22 may be applied to a.c. cables where the electric stress in the insulation is not depending on temperature gradients. The same algorithm can be applied to overload durations shorter than 600 s, to different pre-overload load factors (instead of 0.75), or to aluminum conductors by accordingly adjusting the parameters.

For d.c. cables an additional restriction applies. The temperature difference over the insulation must not exceed a specified value in order to stay below a specified electric stress in the insulation. Equation 3.24a would thus transform for d.c. use:

$$(I^2_{\text{overload}} \cdot R' - I^2_{75\%} \cdot R') \cdot 600\, s = c_p \cdot m_{Cu} \cdot (\Delta_{\max} - \Delta\Theta_{75\%}) \tag{3.24a}$$

with $\Delta_{\max}$ the maximum specified insulation temperature drop, and $\Delta\Theta_{75\%}$ the insulation temperature drop at 75% load.

For longer overload periods, the time-dependent heat outflow through the different layers must be taken into account.

From the situation depicted in curve no. 3 of Fig. 3.6, another overload scenario is also possible. If only a little overload is run, e.g. 15% over full load, the temperature difference over the insulation would soon (within 3–4 times $\tau_c$) assume a new value somewhat larger than the difference at full load. But the conductor temperature would not reach the limit in the first place. Now, we have a non-adiabatic process where the excess heat from 15% overload has time enough to flow outwards through the different layers. It slowly heats up the surrounding soil. As the soil slowly increases in temperature, the conductor temperature keeps pace. When it arrives at the limit (=90°C in Fig. 3.6), the load must be reduced to 100% load to avoid overheating.

Power cables with direct contact to the free sea water (unprotected, laid directly onto the seafloor or in spans) cannot use the benefits of the thermal capacitance of the surrounding soil. The surface temperature of these cables follows the seawater temperature, which is changing only very slowly through the year. The exposed cables will reach steady-state conditions after only a few hours. The overload capability of exposed un-buried cables is much lower compared to buried submarine power cables.

#### 3.1.3.3 Cyclic or Variable Loads

Most submarine power cables are operated with varying load, which is often below the rated power transmission. These cables are under-utilized during certain periods, and the operator might be interested in extra transmission power during peak demand.

Some cables, predominantly those feeding populated islands, may have a daily cyclic load pattern. IEC 60853 has developed algorithms to calculate possible overload due to cyclic cable utilization. The methods are based on the calculation of an average loss factor through the day, such as indicated in Eq. 3.25. The fact that the cables have a daily load pattern can be exploited in two different ways:

1. A higher short-term ampacity can be offered during and past the periods of lower load.
2. A smaller conductor size can be chosen if the load pattern is guaranteed.

Another actual case of variable loads is the submarine connection of OWPs. Most of the time the load in wind-farm cables (both in-field connectors and export cables) is below rated OWP output power. The duration of peak loads is counted in few days at each time. The short-time peak loads are efficiently equalized by the large time constant of the sea soil. A statistical analysis of the expected wind conditions can result in lower cable load requirements and appreciable investment cost reductions [8]. For the unlikely case of long-lasting full-power wind generation, on-line cable temperature monitoring can advise the operator to shut down some turbines if necessary.

However, it must be remembered that unburied submarine cables have no thermal reserve from the ambient soil; they offer very little overload capabilities.

#### 3.1.3.4 Thermal Resistivity of the Seafloor

For good reasons, one of the prime objectives of any cable design is to avoid hotspots as they can jeopardize the availability of the link. It is therefore extremely important to have a detailed knowledge of the thermal surrounding of the cable, i.e. the thermal resistivity of the seafloor and the ambient temperature. Actually, the precise knowledge of these parameters along the cable route can save investment money and/or increase availability and lifetime.

The thermal resistivity of soil is a function of the soil base material, the dry density, the distribution of grain size, the compaction, the humidity and the content of organic materials. The influence of humidity, one of the most important factors for land based soils, can be disregarded in subsea soils because the soil is completely soaked. This is also valid for seafloors of tidal flats which fall dry during low water tide. The German Maritime Authority (BSH, Bundesamt für Seeschiffahrt und Hydrographie) assumes a thermal resistivity of 0.7 K·m/W or less for soil saturated with water [9]. Values up to 1.03 K·m/W have been found in in-situ measurements in the North Sea [10]. Other measurements show values as low as 0.5 K·m/W [11]. Further values are shown in Table 3.6.

The large distribution of these values and those found in Table 3.6 implies that it is one of the most important objectives of the marine survey to measure the thermal resistivity of the soil along the cable route. The measurement of thermal resistivity values for soil is rather delicate. Soil samples taken from the intended installation site can represent the soil base material, grain size distribution, and, in case of submarine soil samples, the humidity content. However, the in-situ degree of compaction is difficult to reproduce in the laboratory. The degree of compaction might be different in virgin soil and in soil after cable laying and burial.

The soil conditions in the vicinity of the submarine cable can be inhomogeneous due to geomorphologic or anthropogenous factors. As an example, the cable is perhaps installed in layered soils or in trenches, which are being filled with non-local material. Also the protection of submarine cables with rock-dumping or concrete slabs creates inhomogeneous thermal ambient conditions. Commercial FEM software can solve the task to find out the effective thermal insulation of an inhomogeneous cable cover that is composed of different soils/rocks/items.

It is important to create a complete picture of the soil conditions along the entire cable route before doing the cable design. The locations of the limited number of soil samples should be chosen so that the thermal properties of the cable route can be mapped in sufficient accuracy.

**Table 3.6** Thermal properties of some submarine soils

| | Heat capacity per volume $\rho \cdot c_p$, MJ/($m^3$·K) | Thermal Resistivity $\rho_T$ | |
|---|---|---|---|
| References | [12] | [12] | [13] |
| Gravel | 2.4 | 0.55 | 0.33–0.5 |
| Sand | 2.2–2.9 | 0.2–0.59 | 0.4–0.67 |
| Clay/silt | 1.6–3.4 | 1.0–2.5 | 0.56–1.11 |

#### 3.1.3.5 Ambient Temperature

The ambient temperature is a critical value in all thermal design calculations, no matter which method is used, or which load cases are considered. The ambient temperature for the cable is defined as the temperature at the *locus* of the cable if the cable would *NOT* be there, i.e. the undisturbed ambient temperature.

For unburied submarine cables, the ambient temperature is simply the temperature of the seafloor water. Even when the surface water temperature may change considerably over the year the water temperature at seafloor level can be quite constant but that is not the case everywhere. At some locations in the North Sea, the seafloor water temperature fluctuates between 8 and 17.5°C during the year. Relevant data for a specific submarine cable project can often be obtained from the national hydrographical institute or commercial survey companies. The highest summer temperature of the seafloor water is different in different years. Statistics list 10-years-high, or 100 years-high values. It is up to the decision of the cable owner/operator which of these values should be used in the context of the overall asset management.

When the unburied cable is carrying load, its surface temperature is only a few degrees over the ambient water temperature. Flowing water provides a better assimilation of temperatures compared to calm water. However, a thermal cable design should not be based on the assumption of the beneficial effects of flowing water.

For the thermal design of buried submarine cables, the ambient temperature at the burial depth must be taken into account. The annual variation of the water temperature at the sea floor penetrates down into the sea bottom until it reaches the intended cable position and beyond. In deeper depths under the seafloor, the amplitude of the seasonal variation becomes smaller and the peak value occurs later in the year. The schematic course of the temperature is shown in Fig. 3.7. Assuming an annual average temperature $T_a$ on the seafloor and a sinusoidal course of the seafloor temperature, the annual variation of the temperature in the depth $z$ can be described as [14]:

$$T(z,t) = T_a + A_0 e^{-z/d} \sin\left(\frac{2\pi(t-t_0)}{365} - \frac{z}{d} - \frac{\pi}{2}\right) \qquad (3.25)$$

$T$ (Z,T) Temperature at time t in the depth z
$T_a$ Average temperature at the seafloor averaged over the year
$A_0$ Amplitude of the annual variation of the seafloor temperature
$d$ penetration depth of the annual temperature variation
$t_0$ Constant reference time.

The penetration depth is $d = (2\ D_h/\omega)^{1/2}$ with $D_h$ being the thermal diffusity (cf. table 3.7). $\omega = 2\pi/365$ is the angular frequency of the annual variation, denoted in 1/day.

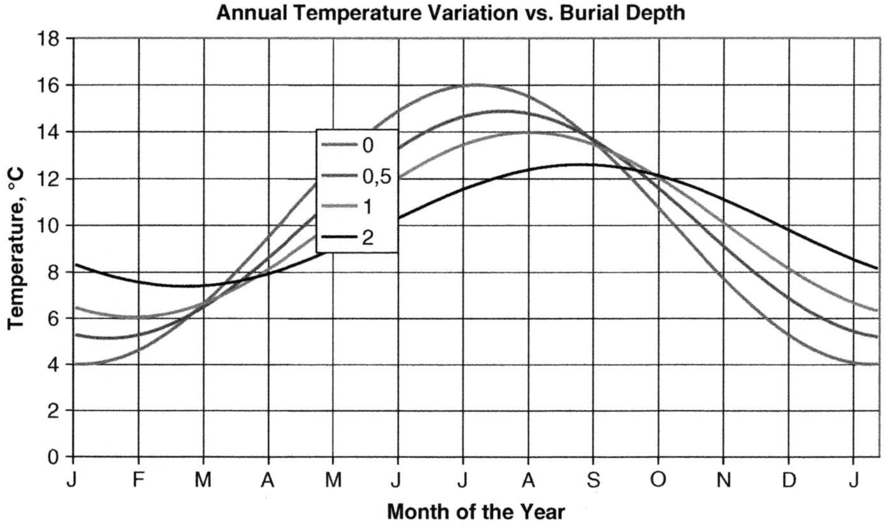


**Fig. 3.7** Schematic course of the temperature as a function of time and depth. The depth parameter is given in *metres*. Assumed thermal diffusity $D_h = 0.05\ \mathrm{m^2/day}$

**Table 3.7** Empirical relationship between thermal diffusivity and thermal resistivity for homogenous and moist soils, according to IEC 60853-2 App. D

| Thermal resistivity (m×K/W) | Thermal diffusivity ($m^2$/day) |
|---|---|
| 0.35 | 0.0864 |
| 0.45–0.54 | 0.0691 |
| 0.55–0.64 | 0.0605 |
| 0.65–0.84 | 0.0518 |
| 0.85–1.04 | 0.0432 |
| 1.05–1.24 | 0.0389 |

The parameters of Eq. 3.25 can be determined by plotting the annual variation of seafloor temperature. From this, the average temperature at the seafloor $T_a$, the amplitude of the annual variation $A_0$, and the reference time $t_0$ can be determined. For all further calculations at various depths the same $t_0$ is used.

In the example of Fig. 3.7, the soil temperature at 1 m burial depth reaches a peak of 14°C. The cable that will be buried here can be designed for 14°C ambient temperature instead of 16°C which is the maximum temperature at the seafloor. Note that Fig. 3.7 is just an example showing the principal course of temperatures. For every submarine power cable project the values $T_a$ and $A_0$ should be identified. If they cannot be retrieved from long-term measurements the estimated values should have sufficient safety margins.

#### 3.1.3.6 Conditions Changing with Time

Seafloor conditions, which have been charted by survey operations, may alter during the cable's lifetime. While water temperature hopefully increases only slowly with the climate change, other parameters may change faster.

Coastal waters exposed to tidal currents are subject to strong and fast changes of the bathymetric structure. Spring tides and storms can cause erratic changes. A cable buried at the –2 m level might be exposed to water, or covered underneath 10 m of sediment. These unpredictable changes have nevertheless to be taken into account for the cable design.

The thermal ambient of submarine cables can also be changed by human activities such as dredging, dumping, etc.

Submarine cables can be subjected to marine growth. The submarine fauna/flora may create extensive layers of organic material over buried cables, which has the effect of thermal insulation and consequential overheating of the cable. Cables on a free span, covered with a thick layer of subsea dwellers, may constitute dangerous hot-spots in the cable route.

### *3.1.4 The 2 K Criterion*

The prospective of a massive installation of offshore wind parks triggered a discussion of the heating of the seafloor. Environmentalists and authorities mainly in Germany advocate a limitation of the expected warming of the seafloor above the submarine power cable (cf. Chap. 10). According to this discussion, the seafloor above the cable must not be warmed up more than 2 K over the undisturbed temperature. The warm-up is calculated for a reference location straight above the cable, at a depth of 0.2 or 0.3 m under the seafloor surface.[9] By means of FEM software the situation can be calculated easily. This chapter provides a simpler method to determine the warm-up for many cases. We use the following terminology.

$\Delta\Theta_{0.2}$ is the temperature rise over the undisturbed sea soil, at a depth of 0.2 m under the seafloor. $\Delta\Theta_{0.3}$ is the corresponding value for a depth of 0.3 m under the seafloor. Both values are valid for steady-state conditions, which are reached only after weeks of continuous constant full load. We consider two representative types of cable installation configurations:

1. A pair of cables laid touching. HVDC cables are sometimes installed in this configuration.
2. A single cable. This represents the case of a three-phase a.c. cable, or one single-core cable in a group of cables laid with distance. In submarine installations, cables are installed either in a bundle (cf. case 1) or with a lateral distance of 5 m or more due to installation restrictions. In the latter case, each cable of the group can be considered as a single cable.

[9]Different authorities and environmental groups use either 0.2 or 0.3 m.

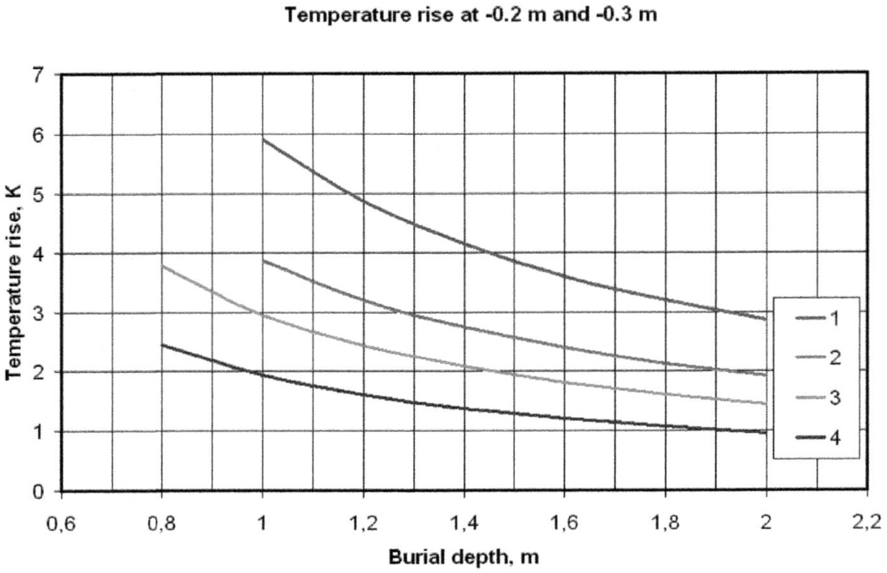


**Fig. 3.8** Temperature rise over the undisturbed seafloor. Curve 1: A pair of HVDC cables, touching, cable diameter 100 mm, –0.3 m; Curve 2: A pair of HVDC cables, touching, cable diameter 100 mm, –0.2 m; Curve 3: A single cable, diameter 150 mm, –0.3 m; Curve 4: A single cable, diameter 150 mm, –0.2 m

The curves in Fig. 3.8 are calculated for a cable loss of $P_{\mathrm{loss}} = 30$ W/m in each cable, and a thermal resistivity of $\rho_T = 1.0$ K·m/W. The temperature rise in a specific project can be calculated as follows:

To find out the temperature increase $\Delta\Theta_{0.2\,\mathrm{m,\,project}}$ at the –0.2 m level for any project the following equation shall be used:

$$\Delta\Theta_{0.2\,\mathrm{m,project}} = \Delta\Theta_{0.2\mathrm{m}} \cdot \rho_T \cdot \frac{P_{\mathrm{loss,project}}}{30\,W/m} \tag{3.26}$$

where $P_{\mathrm{loss,\,project}}$ is the total cable loss in the actual project. $\Delta\Theta_{0.2}$ is taken from Fig. 3.8. The actual thermal soil resistivity goes in as $\rho_T$.

Accordingly, the temperature increase $\Delta\Theta_{0.3\,\mathrm{m,\,project}}$ at the –0.3 m level can be calculated from the values in Fig. 3.8 and the actual cable losses and thermal resistivity.

The cable diameter has only little influence on the resulting temperature rise. A case with a single cable with 190 mm diameter has been calculated. The results differ from the values of the 150 mm cable only in the second decimal place.

The results presented in Fig. 3.8 are strictly spoken valid for steady-state conditions only. However, they can also be used for cases where the cable load is subject to daily cyclic patterns. The large thermal inertia of the seafloor equalizes the daily variations. Starting from the daily load pattern an average cable loss value can be calculated from the mean square $I_{24}$ of the current during the day:

$$P_{\text{loss,average}} = R \cdot s \cdot \frac{1}{24\ \text{h}} \cdot \int_0^{24} I^2(t)\, dt \tag{3.27}$$

where

$I(t)$ the conductor current as a function of time. The integral adds up one current value for each hour for the day.
$R$ resistance of each phase.
$s$ number of phase conductors in the system.

The average loss during a 24-h period is often considerably smaller than the cable losses at full rated load. This should be taken into account when using Eq. 3.26 for the calculating of the soil warm-up.

### 3.1.5 Economic Aspects of the Thermal Design

When designing a power cable it is attempting to reduce the conductor size as much as possible in order to reduce the tender price. This design method results in the smallest conductor meeting the ampacity requirements and the smallest initial costs. Indeed, for longer submarine cable links, the slightest reduction of the conductor size can save appreciable amounts of investment. For a 100 km link the cross section reduction with only 10 mm$^2$ copper can save some 36.000 US$ in copper costs per conductor.[10] However, it became clear decades ago that the costs of losses must not be neglected. A larger conductor cross section reduces the losses and associated costs over many years ahead. A comprehensive cost analysis includes not only the up-front investments (Capex) but also the cost of losses during the cable lifetime. Here, we try to sort out what the loss evaluation factor is, and how it can translate to the conductor size with the lowest lifetime costs for the investor.

The total lifetime costs of a cable system include the investment and the accumulated future costs of losses and maintenance. The future operational costs (Opex) can be projected to the present time with "present value" methods. The question is whether or not the smallest conductor size from the thermal point of view generates the smallest lifetime costs.

The assessment of the present value of future cable losses starts with the calculation of the losses $Q_L$ in one year. The accumulated losses over a year in the cable system are:

$$Q_L = I_0^2 \cdot R \cdot s\mu \cdot 8760\, h \tag{3.28}$$

where

$I_0$ nominal maximum conductor current
$R$ equivalent resistance of each phase

[10] Assumptions for the calculations in this chapter: Copper price 4000 US$/ton.

$s$ number of phase conductors in the system
μ load loss factor of the cable link

The equivalent resistance includes all load-depending losses such as ohmic conductor losses, the contribution of skin and proximity effects, and the contribution of armoring/screen/sheath losses. Dielectric losses are neglected here. The load loss factor $\mu$ is the normalized mean square of the current over time.

$$\mu = \frac{1}{8760\,\text{h}} \cdot \int_0^{8760} \frac{I^2(t)\,dt}{I_0^2} \tag{3.29}$$

*I(t)* is the average conductor current in each hour of the year. For a cable system running on full power over the year $T\mu = 1$. For a cable link to be built in the future, it is not easy to determine the integral in Eq. 3.29. Meanwhile some rough assumptions can be made:

$$\mu = p \cdot m + (1 - p) \cdot m^2 \tag{3.30}$$

with $p = 0.3$ for transmission lines and $p = 0.2$ for distribution lines [2]. $m$ is the load factor, i.e the normalized average current (not square) over time.

We can reduce the future loss costs to the present value $C_L$ at the time of investment:

$$C_L = \frac{C \cdot Q_L}{i} \cdot \left[1 - \left(\frac{1}{(1+i)^n}\right)\right] \tag{3.31}$$

where

*C* Cost of 1 kWh of electricity to cover the losses ($/kWh)
*i* Rate of interest on the capital market
*n* Cable lifetime (years).

This formula can be used both for a.c. and d.c. cables. It is valid under the (uncertain) assumption that *C*, *L*, and *i* are constant over the *n* years of lifetime, which is a strong simplification. Calculation procedures including variable load schemes and variable loss unit costs *C* over the years are given in [2]. With the equations given above the present value $C_L$ of future loss costs can be calculated. This value is often given to the cable manufacturer in the cable system specification in "dollar per kW" units. Each kW of cable losses is penalized with a sum of money reflecting the accumulated $C_L$ costs of the losses over the entire cable lifetime.

Figure 3.9 illustrates the conditions under which it is profitable to increase the copper cross section, depending on the combination of copper price and energy price during the lifetime of the cable. The assumptions are:

- Cable design (1200 $mm^2$ HVDC) according to Table 3.4
- Ampacity 1854 A rated maximum current

- Lifetime 30 years
- 7% capital costs
- Cable price increases linearly with the LME copper price times a factor 1.5 for manufacturing
- The calculation is made for an upward step from 1200 to 1210 $mm^2$.

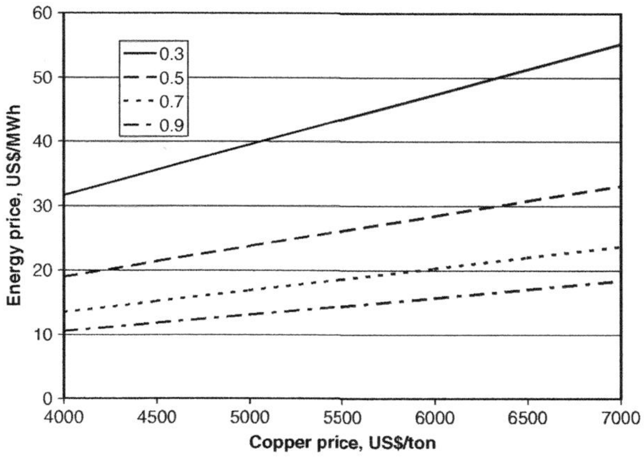


**Fig. 3.9** "Break-even" values for the cable described in Table 3.4. At energy prices over the curve it is profitable to increase the copper cross section

The load loss factor $\mu$ is the curve parameter. It is profitable to increase the copper cross section if the energy price is higher than the value stated in the diagram. The incremental cable price indicated above is nothing but an assumption for incremental cable price changes and must not be taken as an indication for cable prices based on LME copper values.

By nature, all theories are based on assumptions on the future value of some parameters, which tend to be very volatile:

*The technical lifetime* of a cable system can be anything between 10 and 40 years.[11] The lifetime is a statistical number. Most manufactures would agree on a 30–40 years expected lifetime for a well-protected submarine cable.

*Capital market terms* will vary largely during this time, and so will the rate of interest. Since an investor would probably finance the investment with a mix of loans with different terms it is not possible to put a single rate of interests into the formulae. The often used Eq. 3.34 is not very useful in these cases.

*Generation costs for losses* are often equal to the electricity price on the upstream side of the cable link. This price changes by the hour for trading links. For OWP the loss costs can be assessed from different viewpoints as explained in Chap. 10. In any case, the costs for losses are not constant or even known over the cable lifetime.

*Installation criteria.* If there are requirements to keep a certain low sediment temperature over the buried submarine cable, it could be advantageous to reduce

[11] Some submarine cable installations have had lifetimes below 10 years or over 40 years, and may be considered as statistical outliers.

**Table 3.8** Estimated utilization of submarine power cable links

| Type of submarine link | Link utilization |
|---|---|
| Supply of islands with residential demand | According to island demand, 4380/8760 h |
| Supply of G&O platforms | 90% 7884/8760 h |
| OWP | 25% 2190/8760 h |
| Power trading between autonomous grids | 70% 6130/8760 h |

the cable heat losses by extra conductor copper. A larger conductor emits less heat and can be buried shallower for less money to meet the 2 K criteria. At the end, the larger conductor might induce lower costs.

*Cable utilization.* The utilization of a submarine cable link depends largely on the type of the application (cf. Table 3.8). For a specific cable link the utilization may vary largely from week to week and over the years.

*Recycling costs/benefits* after the cable lifetime. Regulators today sometimes require that cable systems must be recovered after the end of life. If this is the case, the conductor copper is not a write-down investment. In contrary, increasing metal prices can turn the cable copper into a valuable asset. Extra copper in the conductor would not only reduce costly losses but also would contribute to late income when recovering the cable.

All these imponderables make the standard economic evaluation methods little useful. Some trends are visible:

- For a short expected lifetime the Capex are more important than the Opex. Select the conductor area as small as possible with respect to thermal conditions.
- For a long expected lifetime the Opex may outweigh the Capex. It might be attractive to put in more copper to reduce costly losses.
- High energy prices bring the focus to the loss reduction. Losses must be paid for somewhere, and the value of some extra copper in the cable becomes more attractive with increasing energy prices.
- An OWP owner might consider investing into a larger conductor size. To bring the same amount of power to shore, it is better to invest in a maintenance-free increase of cable cross section compared to another turbine needing annual maintenance.
- If the cable is to be recovered after life, the investment in extra copper is even more attractive due to future metal prices.

## 3.2 Design of Mechanical Properties

Submarine power cables must be designed to withstand all mechanical stresses during manufacturing, handling, transport, installation, and operation. The stresses imposed to submarine power cables are much different from those imposed to underground cables. An inappropriate mechanical design has the potential to leave the cable vulnerable to damages with the possible consequence of higher unavailability and large repair costs. As a result of poor mechanical design, some cable systems had to be abandoned or replaced by new ones long before their electrical lifetime was over.

The first challenge is to bring the cable safely into the water. The armoring has to provide sufficient tensional strength. The required tensional strength for a submarine power cable project is primarily a function of the water depth. Dynamic forces during the installation may impose much stronger requirements, and conditions during the operation, such as strong currents or forces in cables hanging in free spans, may add to the tensional strength requirements.

### *3.2.1 Tensional Forces During Laying*

When the cable is being laid from a cable ship there are at least four components that contribute to the tensional forces at the laying wheel:

- Static weight of the cable between the laying ship and the seafloor.
- Residual bottom tension, which translates to an extra tensional force at the laying wheel.
- Extra weight of the catenary line between the laying wheel to the touch-down point (TD) on the seafloor.
- Dynamic forces when the laying wheel is moving up and down.

The static tensional force is $T_s = w \cdot D$, with $w$ being the unit weight of the cable in water and $D$ being the water depth. The small length of cable between the water surface and the exit point on the laying wheel[12] is neglected here.

During laying the cable is not just let down vertically. Instead, the cable must be positioned in a well-defined catenary line from the laying wheel to the TD by application of a certain tension in an on-board cable break device. Under these circumstances the cable hits the sea floor asymptotically.

The shape of the catenary is associated to a certain bottom tension. The length of the catenary line is longer than the water depth resulting in a larger weight of cable hanging in the laying wheel as if the cable would hang down vertically (cf. Fig. 7.28). The top tension $T$, i.e. the tension in the cable at the laying wheel, is expressed as:

$$T = \sqrt{T_0^2 + w^2 s^2} \tag{3.32}$$

where $T_0$ is the bottom tension and $s$ is the length of the catenary line. At zero bottom tension, the catenary length becomes $s = D$ and the equation reduces to $T = w \cdot D$.

[12] The expression "laying wheel" stands here also for a laying chute or similar overboarding arrangements.

A complete mathematical treatment of the catenary line is given in the appendix to Chap. 7.

The vertical movement of the laying sheave caused by wave-induced vessel motion adds a dynamic force to the weight of the cable in the catenary line. We assume that the vertical movement of the laying sheave is sinusoidal and find the maximum vertical acceleration $b_{\text{max}}$:

$$b_{\text{max}} = h/2 \cdot (2\pi/P)^2 \tag{3.33}$$

where $h$ is the vertical movement amplitude (measured peak to peak) and $P$ is the movement period (time between subsequent wave peaks). The maximum force on the hanging cable is now

$$T_{\text{max}} = T_s + m \cdot b_{\text{max}} \tag{3.34}$$

where $m$ is the mass of the hanging cable. Since the force due to acceleration is an inertia phenomenon, the cable mass must be used for the calculation rather than the weight in water or air. Each cable laying ship has different sea-keeping characteristics describing how the vessel reacts on different waves from different directions. What matters for the cable top tension, is the vertical acceleration of the laying wheel. Predominantly pitch and heave movements influence the wheel vertical movements. A more detailed treatment of wave statistics and the relation between wave movement and wheel movement is given in Chap. 7.

Real waves sometimes do not obey the sinus equation. Instead, they are, in parts of the wave silhouette, steeper than a sinus curve. In other cases, waves from different directions and different causes can be superimposed, which leads to very steep wave fronts. In these cases the vertical acceleration $b_{\text{max}}$ and hence the dynamic forces may be substantially larger than indicated by the sinus calculation.

The tensional forces on the cable are results of waves with a statistically distributed wave height. Even if the significant wave height of the weather forecast seems to be acceptable, a few waves of exceptional amplitude can cause tensions big enough to damage the cable. The installation engineer must always keep this in mind.

It is evident that the simple sinus representation of the vessel movement is sometimes not adequate to describe statistically rare events. To take into account even unexpected events, guidelines for cargo stowing recommend lashing dimensions fit for vertical acceleration of up to 1 g (=9.81 ms$^{-2}$) at the aft and bow of a vessel ([9], in Chap. 7). If no detailed values about the cable ship's vertical movements are known, a value of $b_{\text{max}} = 6$ ms$^{-2}$ shall be assumed.

It is rather easy to estimate the contribution of the catenary and the wave dynamics to the top tension separately. The results give us an idea on the augmentation of the tensional forces due to both effects. It is much more complex to calculate the combined effects on the tensional forces of a catenary line and the vertical acceleration dynamic.

The entire length of the suspended catenary, also the cable portion close to the TD, must be accelerated vertically. However, the inclined cable can transmit forces only in its own axial direction. For this reason, large tensions along the cable have to be transmitted to create the vertical forces for the cable acceleration close to the TD. A complete laying analysis with resulting tensional and bottom tension values can be performed with specialised commercial software. Some of the software tools also can take the vessel's response to various sea-states into account. Engineering companies working for the offshore industry, and submarine installation companies, often have such tools.

It is a straight-forward task to design the armoring of a submarine power cable for the tensional forces in a catenary line in calm weather. It is more difficult to determine for which sea state the cable should be designed during installation. The selection of the maximum considered sea-state is a matter of attitude rather than engineering. The sea-state statistics for the installation area and the intended installation season can be found out in a desktop study. It must subsequently be decided which range of sea-states the cable should be designed for. If the cable is designed only for the stresses occurring in moderate sea-states, the number of suitable weather windows for installation will be small. If the cable design allows for higher sea-states during installation, more and longer laying opportunities will appear. The costly risk that a laying operation has to be terminated due to adverse weather is reduced.

### *3.2.2 The Cigré Test Recommendation*

The only known mechanical test recommendation for submarine power cables is the Cigré test recommendation for submarine cables [15], referred to as Electra 171 in the following. The most prominent test of the recommendation involves simultaneous bending and pulling of the cable, resembling the stress during pay- out of the cable over the laying wheel. The test arrangement is shown in Fig. 5.9 in Chap. 5. Electra 171 suggests tension values as follows (cf. Fig. 3.10):

For a laying depth up to 500 m:

$$T = 1{,}3 \cdot w \cdot d + H \cdot \tag{3.35}$$

where $w$ and $d$ are as explained above. The additional force $H$ reflects the bottom tension of the cable during installation. $H$ is:

$$H = 0{,}2 \cdot w \cdot d_1 \tag{3.36}$$

with $d_1$ being the laying depth but 200 m at minimum.

For depth greater than 500 m, Electra 171 suggests another tensional force:

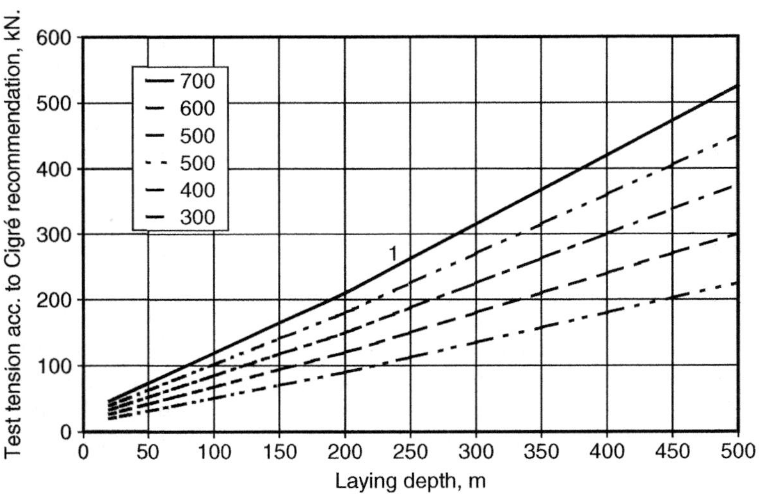


**Fig. 3.10** Tensional force for testing according to the Cigré recommendation. The curve parameter is the weight of the cable in water (N/m)

$$T = 1{,}3 \cdot w \cdot d + H + 1{,}2 \cdot |D| \tag{3.37}$$

where $|D|$ represents the dynamic tensions (in $N$) calculated as:

$$D = \pm{}^{1}/_{2} \cdot b_h \cdot m \cdot d \cdot \varpi^2 \tag{3.38}$$

$b_h$ the vertical movement, crest to crest, of the laying wheel
$m$ mass of cable in air, kg/m
$\varpi$ $2\pi$/t, circular frequency of the vertical movement of the laying wheel, 1/s
$t$ movement period, in seconds.

This formula does not take into account the cable catenary shape but reflects simply the inertial force of the hanging cable under the acceleration during ship movements.

The Electra 171 test values cover most cases of submarine laying operations. In tough weather conditions, however, the resulting tensional forces can be substantially higher. Table 3.9 illustrates this for a typical single core 1200 mm$^2$ extruded HVDC cable. While the Electra 171 test value is 78 kN, expected values in tough weather can reach 106 kN. This value still does not include any allowances for the dynamic forces in the catenary nor any safety margins. With sinusoidal wave shapes a vertical acceleration of 6 ms$^{-2}$ occurs under quite rough sea-states only. However, irregular wave shapes and interfering waves can generate vertical acceleration of this size in the laying wheel at the stern of the vessel.

**Table 3.9** Comparison of tensional test values of Electra 171 and estimated maximum tensional forces in heavy weather

| | |
|---|---|
| Mass of the cable | 29 kg/m |
| Weight in water | 209 N/m |
| Laying depth | 250 m |
| Force attributed to bottom tension ($H$ in Eq. 3.35) | 10.450 N |
| Static force $w \cdot d$ | 52.250 N |
| Tensional test force according to Electra 171 (Eq. 3.35) | 78.375 N |
| Assumed vertical acceleration $b_{max}$ | 6 ms$^{-2}$ |
| Dynamic force $m.b_{max}$ | 43.500 N |
| Total tensional force $T= w \cdot d+ m.b_{max}+H$ | 106.200 N |

### 3.2.3 Distribution of Mechanical Stress Between Conductor and Armoring

In the following, the response of a single-core DWA submarine cable with steel armoring and copper conductor is discussed. The equations can easily be adapted to other armoring and/or conductor materials. When a tension is applied on the cable the elongation and the tensions in the armor wires and the conductor is given by

$$\varepsilon = \frac{F_A}{E_A \cdot A_A} = \frac{F_L}{E_L \cdot A_L} \tag{3.39}$$

$$F = F_A + F_L \tag{3.40}$$

where

$F$ Total tension (N)
$F_A$ Tension in the armoring (N)
$F_L$ Tension in the conductor (N)
$E_A$ Young's modulus of steel (N/mm$^2$)
$E_L$ Young's modulus of the copper conductor (N/mm$^2$)
$A_A$ Total cross section of the armoring wires (mm$^2$)
$A_L$ Conductor cross section (mm$^2$).

The tension in the conductor can be calculated as follows:

$$\frac{F - F_L}{E_A \cdot A_A} = \frac{F_L}{E_L \cdot A_L} \Rightarrow F_L\left(1 + \frac{E_L \cdot A_L}{E_A \cdot A_A}\right) = F \cdot \frac{E_L \cdot A_L}{E_A \cdot A_A} \Rightarrow$$

$$F_L = \frac{E_L \cdot A_L}{E_A \cdot A_A + E_L \cdot A_L} \cdot F \tag{3.41}$$

The conductor stress $\sigma_L$ is as follows:

$$\sigma_L = \frac{E_L \cdot A_L}{A_L\,(E_A \cdot A_A + E_L \cdot A_L)} \cdot F \tag{3.42}$$

The tension in the armoring wires is:

$$F_A = F - F_L \Rightarrow F_A = F \cdot \left(1 - \frac{E_L \cdot A_L}{E_A \cdot A_A + E_L \cdot A_L}\right) \tag{3.43}$$

The total armoring tension is divided between the inner armoring layer ($F_{AI}$) and the outer armoring layer ($F_{AO}$):

$$F_A = F_{AI} + F_{AO} \tag{3.44}$$

$$\frac{F_{AI}}{F_{AO}} = \frac{A_{AI} \cdot E_A}{A_{AO} \cdot E_A} \tag{3.45}$$

where

$A_{AI}$ Total cross section of the inner layer ($mm^2$)
$A_{AO}$ Total cross section of the outer layer ($mm^2$).

Now the tension in the inner and the outer armoring layer can be calculated from Eqs. 3.43, 3.44, and 3.45:

$$F_{AO} = F \cdot \frac{A_{AO}}{A_{AO} + A_{AI}} \cdot \left(1 - \frac{E_L \cdot A_L}{E_A \cdot A_A + E_L \cdot A_L}\right) \tag{3.46}$$

$$F_{AI} = F \cdot \frac{A_{AI}}{A_{AO} + A_{AI}} \cdot \left(1 - \frac{E_L \cdot A_L}{E_A \cdot A_A + E_L \cdot A_L}\right) \tag{3.47}$$

Finally, the mechanical stress in the inner ($\sigma_{AI}$) and outer ($\sigma_{AO}$) armoring layer is:

$$\sigma_{AO} = \frac{F}{A_{AO} \cdot \cos\varphi_O} \cdot \frac{A_{AY}}{A_{AO} + A_{AI}} \cdot \left(1 - \frac{E_L \cdot A_L}{E_A \cdot A_A + E_L \cdot A_L}\right) \tag{3.48}$$

$$\sigma_{AI} = \frac{F}{A_{AI} \cdot \cos\varphi_I} \cdot \frac{A_{AI}}{A_{AO} + A_{AI}} \cdot \left(1 - \frac{E_L \cdot A_L}{E_A \cdot A_A + E_L \cdot A_L}\right) \tag{3.49}$$

$\varphi_O$ is the lay angle of the outer layer (in degrees)
$\varphi_I$ is the lay angle of the inner layer (in degrees).

For many submarine cable projects, in particular in shallow waters, an armoring with thin wires would be sufficient to comply with the tensional bending test according to Electra 171. However, as the maximum number of wires that can be handled in the armoring machine in one run is limited, manufacturers often use thicker wires to achieve a complete coverage of the cable circumference. In doing so, the cable is

**Table 3.10** Guide values for mechanical properties of copper and steel. Steel grade values from European Standard EN 10257-2:1998

| Wire material | Breaking stress (N/mm$^2$) | Yield point (N/mm$^2$) |
|---|---|---|
| Copper wire | 220–240 | 120 |
| Copper wire – welding zone | 170–210 | 70–80 |
| Mild steel (grade 34) | 340 ... 540 | min 210 |
| Steel grade 65 | 650 ... 850 | |
| Steel grade 85 | 850 ... 1050 | |
| Steel grade 105 | 1050 ... 1250 | |
| Steel grade 125 | 1250 ... 1450 | |

equipped with a stronger armoring than the Electra 171 test would require. But the additional steel also provides additional lateral protection, a property much needed in certain projects as the next chapter explains.

In cases where the additional protection of an oversized armoring is not necessary or wanted, some of the steel wires may be replaced by plastic filler wires. This reduces weight, magnetic losses, and possibly costs.

The armoring wire length is longer than the power cable length. Table 3.11 gives the overlength as a function of the lay-length.

One should not conclude that a long lay-length (20 or 25 times the diameter) would save material because of the short overlength in Table 3.11. A long lay-length simply requires more wires to cover the complete cable circumference.

**Table 3.11** Armoring wire overlength as a function of the lay-length

| Lay-length as multiple of the diameter of the armoring layer | Ratio of armoring wire length to cable length |
|---|---|
| 10 | 1.048 |
| 15 | 1.022 |
| 17.5 | 1.017 |
| 20 | 1.012 |
| 25 | 1.008 |

### *3.2.4 Other Forces and Impacts*

The armoring must withstand all forces that can be reasonably expected during installation and operation. The tensional forces that occur during installation can be predicted with a certain degree of accuracy. Other forces and impacts during installation and operation have an accidental nature and hence are indifferent in their character, amplitude and frequency. Many types of external forces and stresses might attack the cable during installation and/or operation:

- Overbending, mostly during installation because of inadequate equipment or insufficient control
- Impact from edges or rocks may happen during burial operations
- Squeezing from cable engines or inadequate roller arrangements
- Impacts from anchors and fishing tackle.

It is difficult to quantify the magnitude of external violence from the accidental events listed here. The design of cable armoring should be based on a compilation of the expected dangers and threats along the cable route, including the hazards arising during the installation. The history and experience from previous submarine cable installations is another valuable source of knowledge. Unfortunately there is no general design rule for the thickness and number of the armoring wires because impacts, threats, and peak tensions are statistic events. Most manufacturers consider the design data as property, not suitable for publishing. Still, there are a few self-evident rules of thumb:

- more steel provides better protection
- harder wires provide a better protection
- double wire armoring is tougher than single wire armoring
- a short-lay rock armoring provides a better protection against lateral impacts at the expense of tensional force.

The optimum armoring wire thickness is depending on several factors. First of all there must be a minimum wire thickness to withstand the external threats such as fishing gear and anchors. On the other hand, most cable manufacturers can apply only a limited number of wires onto the cable. In order to reach a complete cover of wire armoring they must apply thick dimensions of wire for a complete coverage. The thickness of the armoring wires has a large impact on both weight and diameter of the cable. Hence, the cable length in each shipload and the laying schedule are influenced noticeably by the wire thickness.

The American standard ICEA No. S-57-401/NEMA Standards Publication No. WC2 determines the required wire thickness for paper-insulated submarine cables

**Table 3.12** Thickness of armoring wires according to US standard

| Calculated diameter of cable under the armoring bedding mm | Size of armoring wires, mild galvanized steel BWG | mm |
|---|---|---|
| 0–19.05 | 12 | 2.77 |
| 19.08–25.40 | 10 | 3.40 |
| 25.43–43.18 | 8 | 4.19 |
| 43.21–63.50 | 6 | 5.16 |
| 63.53 and larger | 4 | 6.05 |

according to Table 3.12. However, the standard does not tell if this is valid for single or double-layer armoring, or both.

The side-wall pressure impact (SWP) value is often required to know for the planning of the installation. Literally, SWP is the maximum allowed lateral squeezing force, which the cable can stand without serious damage. The name SW-pressure is misleading as it describes a force per unit length (N/m) rather than a pressure (N/m$^2$). The SWP can best be described as the lateral force onto a cable, which is bent around a wheel under a tensional force. SWP can be expressed by:

$$SWP = F_T/R \tag{3.50}$$

where $F$ is the pulling force and $R$ the wheel or bend radius. There are no literature values for the maximum permissible SWP of submarine power cables. However, for some cable projects the tensional force used during tensional bending tests have been reported. The results are summarized in Table 3.13.

**Table 3.13** SWP in some cable projects

| Link | Armoring | Insulation | Test tension kN | SWP kN/m | References |
|---|---|---|---|---|---|
| Morocco – Spain | DWA copper | LPOF | 353 | 70.6 | [16] |
| Italy – Greece | DWA steel | MI | 471 | 94.2 | [16] |
| Gulf of Aqaba | DWA | LPOF | 500 | 100 | [17] |
| Troll A | DWA | Dc XLPE | 375 | 75 | |

It is expected that double-wire armored cables have a substantially higher permissible SWP than single-wire armored cables. A short-lay rock armoring provides probably even higher SWP.

It must be kept in mind that the SWP refers to a distributed lateral force on the cable. The effects of concentrated impacts or lateral forces cannot be treated with the concept of SWP.

Submarine cable producers often are asked for data on the bending stiffness of submarine power cables. A critical parameter for bending stiffness calculations for cables is the friction between the cable layers. Submarine cables are constructed from a variety of different materials. Friction data are known only with limited accuracy, in particular the friction in the armoring layers. For the steel wire armoring layer, coated to a certain extend with bitumen, a complicated mixture of static and dynamic friction coefficients is applicable. Furthermore, the friction and hence the bending stiffness, is strongly depending on the temperature, which influences the bitumen properties, and the amplitude and radius of the bending. For this reason, technical data on the bending stiffness of submarine power cables of standard design are prone to large inaccuracies. The bending stiffness has no influence on the value of the top tension during laying, but can critically change the bending radius at the TD when the laying wheel moves down in heaving.

A few examples of armored cables and their fate might be interesting to know about:

Two mass-impregnated HVDC cables were installed between Denmark and Norway in the mid-70ties. After full-size tests of the resistance of the cable against external aggression, it was equipped with two layers of armoring, 7.0 and 5.6 mm thick, respectively. These cables were put out of service by very heavy fishing gear in 1976 and 1977, and by a dragged towing weight in 1981 [18]. Until 1984, the cables survived the hits from about 30 anchors and fishing trawls which had got entangled with the cables in waters between 160 and 300 m.

The Baltic Cable with two layers 5.0 mm steel wire armoring did survive some violent anchor entanglements but was damaged seriously by the ferryboat "Nils Holgersson" that lost rudder control outside Travemünde/Germany and hit ground and cable. Another outage was caused by an emergency anchoring. At another incident, the cable survived the encounter with a heavy anchor.

Another submarine power cable with two layers of 6.3 mm wires was damaged during the trenching operation, probably by an unsuitable trenching plough.

The notorious Fox Islands Cables between Rockport, Maine, and the islands of North Haven and Vinalhaven, which suffered a large number of mechanical faults (due to anchors, fishing gear, mistreatment during installation and repair, and unsuitable installation over a steep seafloor outcrop) had #4 BWG galvanized wires (=6.054 mm).

### *3.2.5 Vortex Induced Vibrations*

Irregular seafloors and steep underwater slopes can cause free spans of the cable. Submarine cables hanging in free spans are exposed to oscillations when water currents strike them. The water current creates Karmán-vortices, which separate from the cable alternating on the upper and the lower "edge" of the cable. Such vortices can also be observed in the wake of bridge posts in a streaming river. The dispersal of vortices from the cable is called vortex shedding. Each time a vortex is leaving the cable, a force is exerted on the cable. If the cable is horizontal and the water current also is horizontal but perpendicular to the cable, the forces from the vortex shedding point vertically alternatively up and down. The frequency (Hz) of the vortex shedding is

$$f_s = St\frac{u}{D} \tag{3.51}$$

with $f_s$ the vortex shedding frequency (Hz), $u$ the velocity of the streaming water (m/s), $D$ the diameter of the cable (m), and $St$ the Strouhal number [19]. For submarine power cables and relevant current velocities $St$ can be assumed 0.2.

A cable hanging in a free span has a number of natural frequencies with which it can vibrate like a guitar string. The natural frequencies $f_n$ are multiples of the basic natural frequency:

$$f_n = \frac{n}{2} \cdot \sqrt{\frac{T_a}{m' \cdot L^2}} \tag{3.52}$$

**Table 3.14** Basic natural frequency and minimum flow velocity for the establishment of VIV

| Cable span data | Large cable | Small cable |
|---|---|---|
| Cable mass per meter, kg/m | 40 | 20 |
| Length of free span, m | 20 | 40 |
| Tension in the span, kN | 10 | 2 |
| Cable diameter, mm | 110 | 80 |
| Frequency of the basic natural frequency, Hz | 0.40 | 0.125 |
| $V_{\text{min}}$, m/s | 0.22 | 0.05 |

with $n$ the mode number, $T_a$ the tension of the cable, $m'$ the mass per length unit of the cable, and $L$ the length of the free span [20].

The cable in free span is excited by the force from the leaving Karmán vortices with an oscillating frequncy $f_s$. When the exciting frequency $f_s$ is close or equal to one of the natural frequencies $f_n$ the cable may start vibrating in resonance. This phenomenon is called "lock-in". Combining Eqs. 3.51 and 3.52 results in a minimum flow velocity for the lock-in and the establishement of vibrations:

$$u_{\text{min}} = \frac{D}{2 \cdot St} \cdot \sqrt{\frac{T_a}{m' \cdot L^2}} \tag{3.53}$$

According to [20], this simple relation can be used to make a first assessment of the risk for vortex-induced vibrations (VIV) in a given cable installation situation. Table 3.14 shows the calculated results for two typical cable designs.

It should be noted that the diameter of the cable could change drastically by marine growth. The density of marine growth may be set to 1325 kg/m$^3$. If site-specific information is not available the thickness of marine growth can be assumed as indicated in Table 3.15.

As soon as the cable starts vibrating, its natural frequencies are changed. This is partly due to the inertial force of the water, which is pushed by the moving cable. In hydrodynamics, this is addressed by the added-mass coefficient describing a virtual increase of the mass of the accelerating cable due to the dense medium around it (the water). The added-mass coefficient decreases the natural frequencies, more for cables with low own density (such as aluminium cables or three-phase cables) but only little for cables with high density (such as single-core copper cables). Furthermore, the added-mass coefficient is depending on the actual flow velocity [19].

**Table 3.15** Thickness of marine growth on submarine structures [19]

| | Latitude | |
|---|---|---|
| | 56–59° N | 59–72° N |
| Water depth | Thickness of marine growth (mm) | |
| +2 ... –40 m | 100 | 60 |
| Below –40 m | 50 | 30 |

The bending stiffness of the cable may change the natural frequency to higher values. The influence of the bending stiffness can be estimated from the value φ:

$$\varphi = \left(\frac{n\pi}{L}\right)^2 \cdot \frac{E \cdot I}{T_a} \qquad (3.54)$$

where $E{\cdot}I$ is the dynamic bending stiffness of the cable. If $\varphi \ll 1$, the influence of the bending stiffness can be neglected [20].

Once lock-in has established, the vortex shedding frequency is dictated by the actual oscillating frequency of the cable rather than Eq. 3.51.

In long spans, the higher order natural frequencies lie close together. A certain vortex shedding frequency $f_s$ with its associated bandwidth can strike excitation of many natural frequencies or modes simultaneously. However, as the energy included in each mode is small the risk for lock-in is reduced.

An additional complication arises when the flow speed is not uniform along the span of the cable. The response of cables on non-uniform flow fields has been described e.g. in [21, 22]. In the first place, project planners should try to avoid free spans by any means such as route diverting or seafloor leveling.

Some design changes can contribute to reduce the risk of VIV. The cable mass per meter can be changed by a thicker lead sheath, or, in case of three-phase cables, by lead profiles in the interstices between the cable cores [23]. Increased mass-per-length reduces the cable's natural frequency and the minimum necessary flow velocity, but can also reduce the amplitude of the oscillation. VIV suppression strakes can be mounted onto the cable to break up the laminar flow necessary for the vortex creation and shedding. The severity of a given installation situation can be assessed with commercial software packages in use in the offshore oil and gas industry. Reference [19] provides a comprehensive bibliography on the subject.

## 3.3 Electric Design

The electric design of submarine power cables follows the same design principles as those for underground cables. Since submarine cables are often more remote and less accessible for repair, it should be considered to increase the safety margins. The task of this chapter is not to repeat the well-documented principles of the electric design of a.c. power cable [1, 23, 24]. Instead, a very short overview is presented. A little more attention is dedicated to the dielectric properties of d.c. submarine cables as these have not been treated in reference literature to the same extent.

### *3.3.1 The Concept of Electric Strength*

The electric strength of an insulation material is the ability to withstand an applied voltage without a breakdown. If the voltage is higher than the electric strength, an electric breakdown occurs. No matter which mechanism initiates the breakdown, the result is a sudden discharge of the voltage through the insulation. In cable insulation, such an event inevitably leads to a complete failure and an outage of the power line. The cable insulation must be designed such that it can withstand all expected voltages during the lifetime of the cable.

The electric strength is a material property given in kV/mm. There are no fixed values for solid insulation materials, even if tables with electric strength values can be found in many textbooks and journal articles. The electric strength of a given material is depending on a number of parameters under which the electric strength has been measured. Sometimes, the concept of "intrinsic dielectric strength" of a material is used. The expression describes the electric strength of an ultra-pure sample in a laboratory without any deteriorating influences. The published "intrinsic dielectric strength" values of insulation materials are many hundred kV/mm. Those experiments are conducted on very thin layers of insulating material between extremely smooth electrodes. The electric strength of insulation material depends on the volume of the stressed dielectric. Even if the breakdown voltage of a thick layer of insulation material is higher than that of a thin layer, the dielectric strength as expressed in kV/mm is usually smaller. The reason for this is the increased number of impurities and other irregularities in a thick insulation layer compared to a thin layer. These irregularities can act as starting points for a breakdown. For this reason, the electric strength in industrial insulation systems, such as cables, is usually one or two orders of magnitude lower than those that can be achieved in small laboratory materials samples.

Furthermore, the actual dielectric strength is depending on temperature, voltage shape and duration, ageing etc. All these factors should be considered when consulting tabulated dielectric strength values for the electric cable design.

### 3.3.2 The Weibull Distribution

Electric breakdown is a statistic process. Repeating the same experiment with equal samples and test conditions will result in a statistic distribution of results. The Weibull distribution is widely used to describe the results of electric breakdown experiments. The 2-parameter Weibull distribution can be expressed as:

$$f(E) = \frac{\beta}{\eta}\left(\frac{E}{\eta}\right)^{\beta-1} e^{-\left(\frac{E}{\eta}\right)\beta} \tag{3.55}$$

where $f(E)$ is the probability that the sample suffers a breakdown under the application of the electric stress $E$. $\eta$ is called the scale parameter and $\beta$ the shape parameter of the Weibull distribution. Very interesting is the accumulated probability F(E) for a breakdown:

$$F(E) = 1 - e^{-\left(\frac{E}{\eta}\right)\beta} \tag{3.56}$$

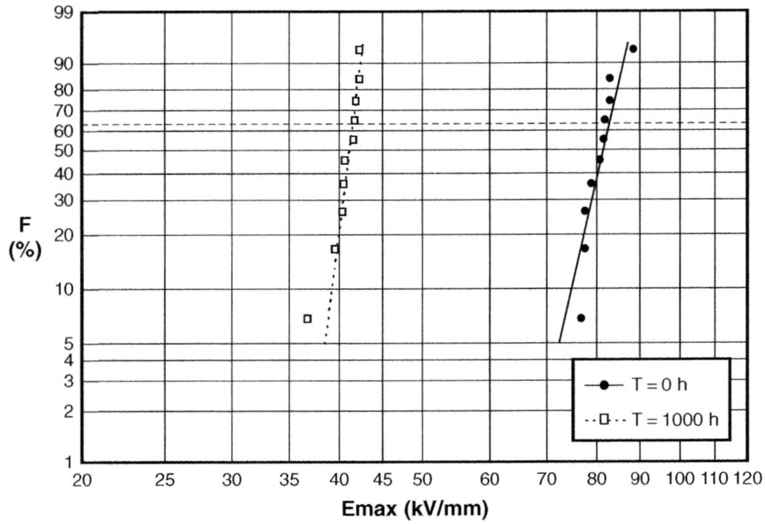


**Fig. 3.11** Results of breakdown experiments plotted on Weibull paper. Explanation in the text (Courtesy of Borealis, Sweden)

For a group of samples, $F(E)$ returns how many of these samples will have failed when tested with an increasing electric stress up to $E$. Figure 3.11 shows a Weibull plot of two experiments with insulation material. The test electric stress is on the x-axis, and the probability of failure is on the y-axis. Plotted on "Weibull paper" with its particular axis formation, Weibull-distributed experimental data should lie on a straight line. From the Weibull plot of experimental data, the scale parameter $\eta$ and the shape parameter $\beta$ can be determined. The scale parameter $\eta$ in Eqs. 3.55 and 3.56 can be interpreted as a measure for the dielectric strength of the material under the given test conditions. The shape parameter $\beta$ tells us something about the statistical scatter of breakdown values over the electric stress scale. A high $\beta$ value indicates a material where the most samples fail within a narrow range of electric stress. A small $\beta$ value indicates a material where samples fail within a large range of stress values. Such a material seems to be less predictable.

The Weibull plot can be used to compare an insulation material in different ageing stages, or different insulation materials. In Fig. 3.11, two sets of breakdown data are plotted into the same diagram. The solid dots represent the results of breakdown experiments on fresh material, while the open dots result from tests on aged material. In spite of the scatter of breakdown voltage as a result of the statistic nature of electric breakdown, it can be seen clearly that the material represented on the right side of the diagram performs better than the left material [25].

Weibull can do more than this. In Eq. 3.56, the variable is the electric stress $E$ and the Weibull distribution describes the failure probability with respect to the applied electric stress. The Weibull statistics can also be used to assess the influence of other parameters on the breakdown behaviour of insulation materials. It is known that the breakdown strength is depending on temperature and the duration of applied voltage. Experiments can be carried out where the breakdown probability is tested as a function of temperature at constant electric stress. The distribution is expected to follow Eq. 3.56 with temperature $\theta$ substituted for the electric stress:

$$F(\Theta) = 1 - e^{-\left(\frac{\theta}{\eta}\right)\beta} \tag{3.57}$$

In this case, *F($\theta$)* describes the accumulated risk of failure for breakdown experiments at temperature $\theta$. The parameters$\eta$ and $\beta$ of this Weibull distribution are naturally different from those for the stress-related Weibull distribution in Eq. 3.56. The parameters can only be established by experiment.

In a similar way, the basic Eq. 3.56 can be used to describe how the electric strength of insulation material depends on the duration of applied electric stress. If $E$ in Eq. 3.56 is substituted with the time duration $T$ of applied voltage, a statistical analysis of the influence of time on the electric strength can be made.

$$F(T) = 1 - e^{-\left(\frac{T}{\eta}\right)\beta} \tag{3.58}$$

Again, *F(T)* describes the probability that a sample fails before or at the time $T$ of voltage exposure. This probability distribution is valid for a certain experiment set-up in terms of temperature, voltage, electrode shape, etc. And again, the parameters$\eta$ and $\beta$ of this Weibull distribution are different from those for the stress-related Weibull distribution in Eq. 3.56.

A utility operating a network of identical cables can plot the cables' lifetime in a Weibull diagram to evaluate the scale parameter $\eta$ and the shape parameter $\beta$. From this evaluation, the mean-time-to-failure (MTTF) of a new cable circuit can be calculated:

$$\bar{T} = \eta \cdot \Gamma\left(\frac{1}{\beta} + 1\right) \tag{3.59}$$

where $\Gamma$ is the gamma function evaluated at ($1/\beta$ +1). The shape parameter $\beta$ can help to estimate if all cables in a circuit will fail approximately during a few years around the MTTF, or if the failures (and repair costs) will be scattered over a longer time period. However, the empirical evaluation of $\eta$ and $\beta$ from the failure data of installed cables is very difficult and probably unreliable. The operation data of the cables such as temperature, magnitude, and duration of overvoltages etc., use to differ so much that no homogeneous cable population can be found for the statistics.

The application of Weibull statistics for power cables is described in detail in [26]. The reference also contains a bibliography on the subject.

Relevant lifetime tests can be made in the laboratory with elevated voltage. The evaluation makes use of the fact that applied stress and lifetime are related:

$$(V - V_0)^n \cdot t = \text{const.} \tag{3.60}$$

where $V$ is the applied voltage (representative for the electric stress, which easily can be deducted from $V$), $V_0$ is a threshold value, and $t$ is the time to failure. The exponent $n$ is a material constant. Test samples from laboratory tests performed under controlled conditions are suitable for statistical treatment [27].

Experimental data can be plotted into a Weibull diagram in order to assess different insulation materials. The Weibull distribution is a valuable tool for researchers to compare different insulation materials in terms of electric strength.

### *3.3.3 Dielectric Design of a.c. Cables*

While the thermal design of a.c. cables is a complex matter due to the extra losses generated by the alternating magnetic field, the electric design of a.c. submarine power cables is so much easier. The cable insulation must be designed in such a manner that the electric stress caused by all expected voltages in the cable system do not exceed the breakdown strength of the insulation, allowing for a safe margin also after reasonable ageing of the insulation.

For a.c. and transient voltages, the stress distribution in the insulation is capacitive. It can be calculated using the Laplace equation $\nabla^2 \Phi=0$ for space-charge free dielectrica, where $\nabla^2$ is the Laplace operator and $\Phi$ the electric potential. In cables, cylindrical coordinates are useful. The Laplacian operator $\nabla^2$ applied to the function $f$ reads in cylindrical coordinates:

$$\nabla^2 f = \frac{1}{r}\frac{\partial}{\partial r}\left(r\frac{\partial f}{\partial r}\right) + \frac{1}{r^2}\frac{\partial^2 f}{\partial \theta^2} + \frac{\partial^2 f}{\partial z^2} \tag{3.61}$$

For the axi-symmetric case of a cable insulation the two last terms of the right side of Eq. 3.61 vanish. After integration of the remaining equation and proper setting of the boundary conditions the electric stress in the a.c. cable insulation can be expressed as:

$$E(r) = \frac{U}{r \ln\left(D_o/D_i\right)} \tag{3.62}$$

with the following symbols:

$U$ applied voltage
$r$ radius in the insulation
$D_o$ Outer diameter of the insulation = insulation diameter under the insulation screen
$D_i$ Inner diameter of the insulation = diameter over the conductor screen.

The applied voltage in Eq. 3.62 is the phase-to-ground voltage $U_0=U_r/\sqrt{3}$ where $U_r$ is the phase-to-phase system voltage. The stress distribution in the insulation of

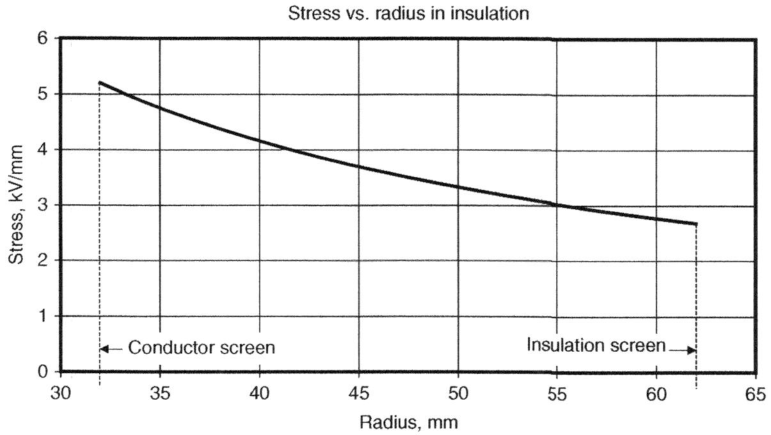


**Fig. 3.12** Electric stress in a 110 kV a.c. XLPE cable

a 110 kV 630 mm$^2$ XLPE cable is shown in Fig. 3.12. The 15 mm insulation wall has $D_o = 62$ mm and $D_i = 32$ mm.

The stress distribution given in Eq. 3.62 and Fig. 3.12 postulates that the relative dielectric constant$\varepsilon_r$ has the same value in the entire insulation.$\varepsilon_r$ is practically independent on temperature and power frequency. Hence the stress distribution formula 3.62 is valid for all a.c. cables, no matter if it is a three-core cable, a single core cable, cold or loaded.[13] The concentration of electric stress at the conductor screen imposes high requirements on material cleanliness and a smooth interface without defects.

Equation 3.62 is also valid for the calculation of stresses under transient voltages such as lightning or switching impulses but must not be used for the calculation of stresses in d.c. cables.

Equation 3.62 can also be used to determine the required insulation thickness for the cable to withstand the expected voltages. The crucial task is to establish the highest acceptable stress for power frequency voltage, switching impulse and lightning impulse taking into account system overvoltages.

#### 3.3.3.1 Overvoltages

During testing and operation of submarine power cables a number of different voltage shapes can occur. Power frequency overvoltages can occur during a single-phase-to-ground cable failure until the fault is cleared. Depending on the system earthing concept, the overvoltage over the insulation of the healthy phases can be

[13] Only some low-voltage cables without insulation screens maintain non-circular stress distributions.

3 times the rated phase-to-ground voltage. This overvoltage, though not very long lasting, can result in extra ageing or even breakdown.

Other overvoltages can be caused by abnormal operation conditions such as incorrect generation/load balance, especially when the submarine cable is used to connect a remote isolated load or generator to the shore grid. A detailed system study can clarify the expected overvoltages in the particular project, and suggest suitable protection measures.

With respect to the strong correlation between overvoltages and ageing, it should be avoided to operate submarine power cables at the maximum design voltage $U_m$ during the greater part of the operational time.

Voltage impulses with high magnitude and very short duration may occur in power transmission systems as a result of switching operations and atmospheric lightning. The duration of these voltage peaks is in the range of micro- or a few milliseconds. The electric stress in the insulation during impulse voltages can be calculated using Eq. 3.62. Fortunately, the dielectric strength of insulation materials under impulse stress is much higher than under power frequency voltage.

Test standards recommend test voltage level for Lightning Impulse tests (LIWL) and Switching Impulse tests (SIWL), cf. Table 3.17. Test levels for a.c. routine tests are summarized in Table 5.5.

#### 3.3.3.2 Design Rules

Scientific books and research articles can provide dielectric strength values for different materials and experimental conditions but they can hardly provide design rules. In addition to laboratory results, the experience and safety policy of power utilities influence the selection of safety factors for the dielectric cable design. National or international standard committees, power utilities and each cable manufacturer have developed own attitudes concerning safety factors and design rules. Given the "minority" role of submarine power cables compared to land-based power cables, almost all cable standards refer to underground cables, leaving the field of submarine cables with little regulation.

The USA industrial standard AEIC CS9-06 lists maximum stress values and the corresponding insulation thickness (Table 3.16)

The European International Electric Commission (IEC) does not recommend any particular insulation wall thickness for high-voltage cables. Instead, test procedures both for power frequency and impulse withstand tests are specified, and a cable design is approved if the relevant type tests are passed (cf. Chap. 5).

Power utilities may have their own cable design standards. Major German TSO's have agreed on a common standard for the insulation thickness of 110 kV XLPE cables: 18 mm. Most cable manufacturers would support a thinner insulation such as 15 mm.

For medium-voltage cables ($\leq$ 36 kV) with an insulation thickness of 4–8 mm, the maximum design stress in the insulation is only between 2 and 4 kV/mm. Modern XLPE insulation has a much higher breakdown strength than this. However, the thin insulation wall renders this insulation more sensitive to production irreg-

**Table 3.16** Stress limits and insulation thickness for extruded cables according to [28]

| Rated voltage (kV) | Conductor size, ($mm^2$) | Nominal internal ac stress limit (KV/mm) | Corresponding generic insulation thickness (mm) |
|---|---|---|---|
| 69 wet | 240–2000 | 4 | 16.5 |
| 69 dry | 240–2000 | 6 | 12.0 |
| 115 | 400–2000 | 8 | 15.0 |
| 138 | 400–2000 | 8 | 18.0 |
| 161 | 400–2000 | 9 | 20.0 |
| 230 | 500–2500 | 11 | 23.0 |
| 345 | 500–2500 | 14 | 26.0 |

Note. The nominal internal stress is the electric stress at the conductor screen at rated phase-to-ground voltage. A "wet" design is a cable without impermeable metallic sheath. The limits given in the standard are valid unless proposed otherwise by the manufacturer or purchaser.

ularities. Higher design stress and thinner insulation would be possible but would require a more definite and costly production and the use of higher grades of insulation material.

For higher rated cables ($\geq$ 150 kV), higher design stresses are generally accepted as a result of better polymer grades and a higher relative thickness stability of the insulation. This can be achieved despite the fact that the specific breakdown strength in thicker XLPE layers is lower than in thin XLPE layers [29].

### 3.3.4 Dielectric Design of d.c. Cables

Direct current high-voltage power transmission is more than a century old, and paper-insulated submarine cables have been used for submarine d.c. transmission since more than 50 years [30]. Extruded d.c. cables have been introduced commercially more than a decade ago by ABB and are now used as submarine cables in a large scale. The increasing need for long-haul submarine power transmission fuels a large interest in the properties of d.c. cable insulation materials.

The d.c. voltage of normal operation creates a field distribution in the cable dielectric that is controlled by the specific conductivity $\sigma$ of the insulation material. The specific conductivity $\sigma = \sigma(E, T)$ is a function of local electric field and temperature. For this reason, a simple analytic solution of the Laplace equation, as in the case of a.c. cables, cannot be applied.

Various expressions for the dependence on $E$ and $T$ of $\sigma(E, T)$ have been published, such as the following for polymers:

$$\sigma\ (E{,}T) = A \exp\left(\frac{-\phi \cdot q}{k_B T}\right) \frac{\sinh\left(B\,|E|\right.}{|E|} \tag{3.63}$$

where $A$ and $B$ are constants, $\phi$ is the thermal activation energy in eV, $q$ is the elementary charge, $T$ the temperature in Kelvins and $E$ the electric field in V/m

[31]. According to other references $\sigma(E, T)$ can be expressed as:

$$\sigma = \sigma_0 \cdot e^{(\alpha(T-273)+\gamma|E|)} \tag{3.64}$$

where $\sigma_0$ is the conductivity at 0°C and 0 kV/mm, $\alpha$ is the temperature dependency coefficient, $\gamma$ is the field dependency coefficient, and $E$ is the local electric field strength [32]. For mass-impregnated cables, we can assume the following:

$$\sigma_0 = 1 \times 10^{-16}\Omega^{-1}\mathrm{m}^{-1}, \quad \alpha = 0.1\,K^{-1}, \quad \gamma = 0.03\,\mathrm{mm/kV}.$$

It is difficult to measure the dependence of $\sigma$ of $E$ and $T$ as the probes usually need a very long time to stabilize.

Although Eqs. 3.63 and 3.64 are different, they describe two fundamental properties of cylinder-symmetrical d.c. insulation:

1. The higher the electric field at a particular radius $r$ in the insulation, the higher the local conductivity $\sigma(r)$.
2. The higher the temperature at a particular radius $r$ in the insulation, the higher the local conductivity $\sigma(r)$.

These properties have far-reaching consequences. Consider a d.c. cable which has just been put under d.c. voltage. In the first instant, the field distribution is according to the green dotted line in Fig. 3.13. The distribution resembles that of an a.c. cable such as shown in Fig. 3.12, with the highest stress at the conductor screen. The relatively high electric field close to the conductor results in a higher specific conductivity in the innermost cylindrical shell of the insulation. As a result, the voltage drop over the shell and the gradient in the same place decrease. After a while, the cold cable d.c. stress distribution has established. This process is called relaxation, and the resulting stress distribution is depicted in Fig. 3.13 as the blue solid line. The distribution represents the situation in a not-loaded cable wich has been under operational d.c. voltage for some hours. The positive effect of the field depending conductivity is to reduce the stress where it is the highest, i.e. next to the conductor. How much the stress peak at the conductor screen is reduced, is depending on the $\sigma$(E) function, which is slightly different for different materials.

The next step would be to switch on the current. Once conductor current is flowing, a temperature gradient over the insulation is established, and the situation changes dramatically. As the temperature rises in the insulation close to the conductor, the local conductivity increases further; and this in turn decreases the local stress. The influence of the temperature gradient is much stronger than the influence of the electric field previously was. As a result the stress on the conductor screen decreases considerably, while the stress close to the insulation screen accordingly increases. With large temperature gradients over the cable insulation, the stress profile can actually reverse leaving the cable with the highest stress at the insulation screen. Figure 3.13 shows the electric stress in a d.c. insulation also for full power (red dashed curve) and fully developed thermal gradient.

To describe the d.c. stress distribution the Laplace equation $\nabla^2 \Phi = 0$ of the a.c. distribution is replaced by the more general Poisson equation:

$$\nabla^2 \Phi = \rho \cdot \varepsilon_r$$

where $\rho$ is the local space charge density. Now the insulation layer is no longer space-charge free as in the case of a.c. cable insulation. The transition between the initial transient stress distribution, as depicted by the green dotted line to the no-load d.c. distribution (blue solid) and to the full-load d.c. distribution (red dashed), is associated with the creation and relocation of space charges in the insulation to fulfil Poisson's equation. As space charges have a limited mobility it takes some time to establish a stable stress distribution (a process sometimes called relaxation). The charge mobility is strongly depending on the temperature [33, 34] so that the relaxation speed is different in cold and warm cables. In d.c. submarine cables at 6°C (the seafloor temperature often found in waters in moderate climate zones), it can take many hours to achieve a stable stress distribution. For the design of submarine d.c. cables, it is important to calculate the time-dependant stress distribution under the influence of all expected operational conditions.

The curves of Fig. 3.13 depict the global stress distribution in the d.c. cable insulation. However, this is only a part of the total picture. Many measurements have demonstrated that space charges also accumulate in the immediate vicinity of the dielectric surfaces of the insulation, i.e. close to the extruded conductor screen and close to the extruded insulation screen. These space charges can be caused by charge injection from the semi-conducting surfaces. Sometimes they are homo-charges (same polarity as the neighbouring semi-conductor), sometimes hetero-charges. This space-charge accumulation has led to many failures in the develop-

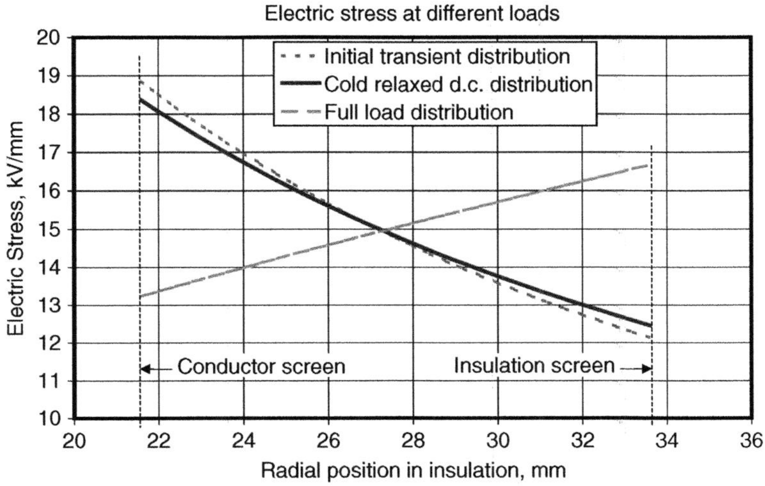


**Fig. 3.13** Electric stress in d.c. cable insulation under different load conditions

ment of extruded d.c. cable insulation systems. Only by orchestrated properties of the insulation material and the screen material the problems of space charge related field distortion can be overcome. Today, at least one manufacturer has succeeded and installed more than thousand of kilometres of extruded HVDC cable.

A comprehensive bibliography on space charge phenomenon in polymeric d.c. cables is given in [35].

### 3.3.5 Dielectric Design of Mass-Impregnated Cables

For practical considerations, the steady-state stress distribution in a mass-impregnated d.c. cable can be calculated analytically according to [36]:

$$\beta = \frac{\alpha \cdot W_C}{2\pi\lambda} \qquad \gamma = \frac{k \cdot U}{r_S - r_C} \qquad \delta = \frac{\beta+\gamma}{\gamma+1}$$

$$E(r) = \frac{\delta \cdot U}{r_S} \cdot \frac{\left(\frac{r}{r_S}\right)^{\delta-1}}{1-\left(\frac{r_C}{r_S}\right)^{\delta}} \quad \text{kV/mm} \tag{3.65}$$

with the following symbols (partly different from [36]):

$\alpha$ [1/K] = conductivity dependence of the temperature
$k$ [mm/kV] = conductivity dependence of the electric stress
$U$ [kV] = voltage conductor – sheath (screen)
$r_c$ [mm] = radius of conductor screen
$r_s$ [mm] = radius of insulation
$r$ [mm] = radius where $E$ is calculated
$W_c$ [W/m] = conductor losses
$\lambda$ [W/K×m] = thermal conductivity of insulation.

The factor $\beta$ may also be written:

$$\beta = \frac{\alpha}{2\pi\lambda} \cdot W_C = \frac{\alpha}{2\pi\lambda} \cdot \frac{\Delta\theta \cdot \lambda \cdot 2\pi}{\ln \frac{r_S}{r_C}} = \frac{\alpha \cdot \Delta\theta}{\ln \frac{r_S}{r_C}}$$

The dielectric behaviour of mass-impregnated cables is governed by a particular effect. In contrary to an oil-filled cable, which has an external pressure supply, the mass-impregnate cable has no inner pressure when it is not carrying a load current. Small voids exist in the gaps between the paper tapes, limited in the radial extension by the thickness of the adjacent paper tape. When the cable is loaded with transmission current, the expanding impregnation compound fills up all voids and creates finally an overpressure inside the insulation. Under full load, the dielectric strength of the cable insulation is much higher compared to the no-load situation. For this reason, the cable could be operated at higher voltage under full load conditions compared to no-load conditions. Experience has shown that the dielectric strength undergoes a minimum during the first one-three hours after reducing the current and, hence, the temperature.

The relevant type tests for mass-impregnated d.c. cables stipulate load cycles with 8 h loading and 16 h cooling ( [7], in Chap. 5). The recommended test voltage during the entire load cycle test is $1.8 \times U_0$. According to experience, the cooling phase of the load cycle is the limiting challenge for a given cable design. In practice the mass-impregnated submarine d.c. cable is designed to fulfil the requirements of the Electra 171 type test requirements rather than the challenges of real cable life.

Initially the rated voltage of the tested cable is set such that the cooling phase of the type test will be passed. Doing so, the additional dielectric strength during the full load period cannot be used for real cable operation. In order to utilize the cable in a more economic way, a voltage regulating algorithm has been installed in some HVDC links, which increases the operational voltage on the cable whenever possible using the higher dielectric strength of the insulation under load conditions. The continuous transmission power can be increased by about 25% using this voltage regulation ("CDVC").

Historically, an operational dielectric stress of up to 40 kV/mm has been considered possible for mass-impregnated d.c. insulation [37]. Today, a d.c. design stress of 25–35 kV/mm has been established. The design impulse strength is usually in the range 80–95 kV/mm for impulse tests superimposed on a d.c. voltage with opposite polarity, as stipulated by the Cigré test recommendation. It should be mentioned that the required impulse test levels of mass-impregnated submarine d.c. cables could be lowered noticeable if converter station manufacturers and cable manufactures agreed on closer protection levels for the cable terminations.

### *3.3.6 Impulse Stress*

For all cable types, there are test standards to demonstrate the impulse withstand level. Often, a Switching Impulse Withstand Level (SIWL) and a Lightning Impulse Withstand Level (LIWL) are defined. Table 3.17 lists test levels for LIWL and SIWL

The electric cable cores of submarine power cables are developed and tested according to the same principles as land cables. Therefore, they have to comply with the impulse test requirements given in the applicable land cable standards. Knowing the impulse test voltage, the electric stress in the cable during the test can easily be calculated with Eq. 3.62.

The design impulse stress of power cables can unfortunately not be found in tables. Impulse breakdown voltages are strongly depending on sample size, material purity, manufacturing quality, temperature, electrode structure, etc. Paper insulation tends to have a narrower statistical distribution of breakdown voltage than extruded insulation. For mass-impregnated HVDC cables Electra 189 requires the same peak voltage ($=2.0 \times U_0$) for both Lightning (LI) and Switching Impulse (SI) tests, while the SI test level is lower than the LI level in IEC test standards. In real life, HVDC

**Table 3.17** Impulse test levels for a.c. cables

| Rated voltage Kv a.c. | Lightning impulse test voltage, kV | Switching impulse test voltage, kV |
|---|---|---|
| According to IEC 60840 | | |
| 20 | 125 | 42 |
| 30 | 170 | 63 |
| 45 | 250 | 65 |
| 66 | 325 | 90 |
| 110 | 550 | 160 |
| 132 | 650 | 190 |
| 150 | 750 | 218 |
| 220 | 1050 | 318 |
| 275 | 1050 | 400 |
| 330 | 1175 | 420 |
| 400 | 1425 | 440 |
| According to British Electricity Boards Engineering Recommendations | | |
| 76/132 | 640 | 380 |
| 160/275 | 1050 | 850 |
| 230/400 | 1425 | 1050 |

cables of any kind would hardly ever experience overvoltages of the magnitude tested in IEC impulse tests.

Extruded d.c. cables for VSC use are terminated indoors and do not suffer any lightning overvoltage in the classic sense at all. Neither the classic switching overvoltage is experienced for these cables.

### 3.3.7 Availability and Reliability

The choice of the insulation wall thickness is as much an asset management task as it is an engineering task. Considering the Weibull plots of the cable insulation material, the cable designer will chose an insulation wall thickness which guarantees that the probability of a cable breakdown within the required life time of the cable is below an acceptable level, given the operation temperature and the electric stress on the cable. It must be noted that the statistical risk for breakdown cannot be eliminated completely unless the cable is not being used.

Given the difficulty of submarine cable installations/repairs, it can be prudent to apply conservative design rules for submarine power cables. The large expenses for submarine cable repair might motivate the use of thicker an insulation than it is used for land-based cables, with the purpose of reducing the electric stress and hence the statistic risk of electric failure. On the other hand, experience from submarine power cable operation demonstrates that there are almost no spontaneous electric failures in submarine power cables (cf. Chap. 9).

A thicker insulation would also increase the costs for lead and steel. In the worst case, the larger cable weight and volume would require a larger cable ship or more laying campaigns.

Rather than adding extra insulation the cable should be protected against overvoltage and excess temperature, in order to comply with the rules of thumb (cf. Chap. 2):

- Increase the operating temperature by 8–10°C and you'll cut the lifetime by half.
- Increase the operating voltage by 8–10% and you'll cut the lifetime by half.

## References


1. Heinhold L (Ed.) (1990). Power Cables and their Applications, Siemens Aktiengesellschaft 1990, Berlin/Munich, Germany, ISBN 3-8009-1535-9.
2. Anders G J (1997). Rating of Electric Power Cables, IEEE Press, McGraw-Hill, 1997, ISBN 07803-1177-9.
3. Brakelmann H (1985). Belastbarkeiten der Energiekabel, VDE-Verlag, 1985, ISBN 3-8007-1406-X.
4. IEC Standard 60287 (2006). Calculation of the Continuous Current Rating of Cables, International Electrotechnical Commission, Geneva, Switzerland.
5. Neher J H (1964). The Transient Temperature Rise of Buried Cable Systems, IEEE Transactions on Power Apparatus Systems, Vol. PAS-83, pp. 1345–1351.
6. McGrath M H (1964). Discussion Contribution to Neher, IEEE Transactions on Power Apparatus Systems, Vol. PAS-83, p. 113.
7. George F Moore (1997). Electric cables handbook, Blackwell Science, ISBN 9780632040759.
8. Brakelmann H (2004). Bemessung der Energiekabel zur Netzanbindung von Offshore-Windfarmen, Energiewirtschaft Vol. 4, No. 103, (in German).
9. BSH: Standard (2007). Konstruktive Ausführung von Offshore-Windenergieanlagen. Hambnurg und Rostock 2007. BSH-Nr. 7005 (in German).
10. Newson T A and Brunning P (2004). Thermal Conductivity of Deepwater Offshore Sediments, International Journal of Offshore and Polar Engineering, Vol. 14, No. 4, December 2004.
11. Wright J F et al. (2002). Thermal Conductivity of Sediments ...., Geological Survey of Canada, Bulletin 585, 2002.
12. VDI Richtlinie 4640, Blatt 1 (2000). Thermische Nutzung des Untergrundes, Table 1. Beuth Verlag, Berlin, Germany.
13. Smolczyk U (2001). Grundbau Taschenbuch Teil 2, Geotechnische Verfahren: Anhaltswerte zur Wärmeleitfähigkeit wassergesättigter Böden, Ernst & Sohn-Verlag, Berlin.
14. Hillel D (1982). Introduction to Soil Physics, Academic Press, San Diego, CA, USA.
15. Cigré WG 21.02 (1997). Recommendations for Mechanical Tests on Sub-Marine Cables, Electra No. 171, April 1997.
16. Valenza D and Cipollini G (1995). HVDC Submarine Power Cable Systems – State of the Art and Future Developments, International Conference on Energy Management and Power Delivery, 1995. Proceedings of EMPD '95.
17. Balog G et al. (1998). The Gulf of Aqaba Submarine Cable Link. Cigré paper 21-302, Cigré Session 1998.
18. Hauge O et al. (1988) Performance of the ± 250 kV HVDC Skagerrak Submarine Cables, IEEE Transactions Power Delivery, Vol. 3, No.1, January 1988.
19. Environmental Conditions and Environmental Loads, Recommended Practice DNV-RP-C205 issued by Det Norske Veritas, Norway, April 2007, Amended April 2008.

20. Balog G et al. (2006). Vortex Induced Vibration on Submarine Cables, Cigré Paper B1-208, Cigré Session 2006.
21. Vandiver J K et al. (1988). Predicted and Measured Response of flexible Cylinders in sheared Flow, Proceedings ASME Winter Annual Meeting, Symposium on Flow-Induced Vibration, Chicago, December 1988.
22. Kim Y H et al. (1985). Vortex-Induced Vibration and Drag Coefficients of Long Cables Subjected to Sheared Flows. Proceedings of Fourth International Offshore Mechanics and Arctic Engineering Symposium, ASME, Dallas, TX, February 17–21, 1985.
23. McAllister D (1982). Electric Cables Handbook, Granada Technical Books Ltd, 1982, ISBN 0-246-11467-3.
24. Bartnikas R (Ed.) (2000). Power and Communication Cables, IEEE Press Series on Power Engineering, ISBN 0-7803-1196-5.
25. Personal Communication Ulf Nilsson, Borealis, Sweden.
26. Bernard G (1989). Application of Weibull Distribution to the Study of Power Cable Insulation, Electra No.127, December 1989.
27. Chan J C et al. (1988). Comparative Performances of EPDM Versus XLPE as Insulation for Distribution Cables, RGE-No. 3-Mars 1988.
28. AEIC Cs9-06 (2006). Specifications for Extruded Insulation Power Cables and Their Accessories Rated above 46 KV through 345 KV (1st Edition). Association of Edison Illuminating Companies, USA.
29. Beyer M et al. (1986). Hochspannungstechnik, Springer, Berlin, Heidelberg, New York, London, Paris, Tokyo, 1986, ISBN 3-540-16014-0.
30. Worzyk T (2007). 100 Years of High Voltage DC Links, Modern Power Systems, Vol. 27, No. 11, pp. 21–24.
31. Boggs S et al. (2001). Effect of Insulation Properties on the Field Grading of Solid Dielectric DC Cable, IEEE Transactions on Power Delivery, Vol. 16, No. 4, pp. 456ff, October 2001.
32. Jeroense M J P (1997). Charges and Discharges in HVDC Cables: in Particular in Mass-Impregnated HVDC Cables, Ph.D. Thesis, Delft University of Technology, The Netherlands, 1997, also published as ISBN 90-407-1438-X, Delft University Press.
33. Bambery K et al. (1996). The Temperature Dependence of Space Charge Accumulation in Crosslinked Polyethylene, IEEE International Symposium Electrets, pp. 16–21.
34. Lim F N et al. (1999). The Temperature Dependence of Space Charge Accumulation and dc Current in XLPE Power Cable Insulation, IEEE Conference on Electrical Insulation Dielectric Phenomena (CEIDP), pp. 66–69.
35. Chong Y L et al. (2007). Temperature Effect on Space Charge Dynamics in XLPE Insulation, IEEE Transactions on Dielectrics and Electrical Insulation, Vol. 14, No. 1, pp. 65–76, February 2007.
36. Eoll C K (1975). Theory of Stress Distribution in Insulation of High-Voltage DC Cables: Part 1, IEEE Transaction on Electrical Insulation, Vol. Ei-10, No. 1, pp. 27–35, March 1975.
37. Schjölberg-Henriksen (1933). Transmission d'énergie électrique par courant continu à de très haute tensions. Berichte der Hochspannungskonferenz, Paris, 1933 (in French).

# Chapter 4
# Accessories

## Contents



## 4.1 Submarine Cable Joints

The manufacture of a submarine power cable joint[1] requires valuable vessel time, specialised equipment for the manufacturing and deployment of the joint, highly specialized teams on board, and a sufficient long period of suitable weather conditions. The failure of some early installation joints during service shaded the reputation of submarine power cable joints. Failures in the joints were usually caused by poor engineering or inadequate installation procedures. As an example, all shore joints in the first Cook Strait cable connection failed or needed repair. The failures were blamed to inadequate engineering of the transition joints connecting different conductor sizes [1]. However, submarine power cables of today deserve a better

[1] The expression "splice" is sometimes used.


DOI 10.1007/978-3-642-01270-9_4, 

reputation. In a 1986 Cigré study on the reliability of submarine power cables, the joints account for 18% of the failures, the cables for 82%. Improved engineering methods, better route survey methods, and more sophisticated installation procedures have changed the picture during the past twenty years. With these improvements, the major submarine power cable manufacturers can provide joints of high quality and reliability. Assembled by well-trained teams and supported by adequate vessel equipment, the joints meet all requirements and do not have to be regarded as a weak point anymore. Still it is prudent to reduce the number of cable joints as much as possible because the joint assembly requires extra good weather time.

Submarine power cable joints come in a row of different shapes, which can cause some confusion. There are factory joints, installation joints, repair joints, flexible and stiff joints, both for 1C and 3C cables. The following paragraphs try to eradicate some of the confusion.

### *4.1.1 Factory Joints*

A factory joint connects semi-finished pieces of cable before the armoring is applied. Factory joints are also used when production mishaps require that the production cable length must be cut to remove damaged parts. The flexibility of the joint allows for application of a continuous armoring over the joint in the ordinary armoring machine in the factory.

The manufacturing of a factory joint (cf. Fig. 5.2) starts with the connection of the conductors of the two cable ends. Different welding methods such as TIG, MIG, etc. are used. Friction welding can be used to connect aluminium to copper. Stranded conductors are welded either with a single welding seam across the entire diameter, or wire-by-wire. The choice of an appropriate welding process and suitable filler materials is critically important to producing conductor joint with sufficient mechanical properties. Welding defects such as discontinuities, cracks, porosity, incomplete fusion or penetration, and nonmetallic inclusions must be avoided. Sometimes, the weld quality is checked by X-rays. The tensile strength of the welded joint is of great importance for the installation of submarine power cables. The conductor portion adjacent to the welding is weaker because it is annealed by the welding heat. The strength is often only 70% of the strength of the original conductor. Less frequently, other methods of conductor joints are used, such as flush ferrules to be crimped on the conductor ends, or soldering processes [2].

The electric conductivity of the weld must be sufficiently high to avoid a hot-spot in the cable. Still, a local moderate increase of the specific resistivity of the weld materials is uncritical as the heat generated from the excess losses is dissipated efficiently into the adjacent conductor.

Screw connectors are not used for factory joints because they would add up to the diameter and obstruct further manufacturing.

After the conductor joint, the insulation is built up, normally comprising the same structure as the cable insulation. The insulation of the two cable ends is tapered to form conical surfaces. New insulation material is applied between the two tapered cable ends now. A longer interface cone provides a lower axial electric field along the sensitive interface. In most cases, the joint insulation is somewhat thicker than the cable insulation in order to reduce the electric stress.

Factory joints according to the described generic design can be used both for paper-insulated and polymeric cables.

The joint insulation for polymeric insulated cables (XLPE, PE, EPR) is made from tapes of similar material, which are lapped in the gap between the cables. The screens are made from carbon-black loaded polymeric tapes. The joint insulation is cured under heat and pressure so that the tapes melt together and form a homogeneous continuous material without voids. If XLPE tapes are used, the curing time is longer for the cross-linking process. The applied pressure helps to suppress the formation of gas bubbles when the XLPE tape is cross-linking. The interface between cable insulation and joint isolation must not have any voids, gaps, cracks, or contaminations. The preparation of the conductor screen and the transition between the conductor screen in the cable and the screen in the joint is a delicate task and requires the highest carefulness (cf. Fig. 4.1). The adhesion of the joint insulation to the cable insulation is of great importance for the electric strength of the joint.

For voltages up to and including 110 kV also self-amalgamating tapes can be used as joint insulation. These tapes must be applied with well-defined tension. The self-amalgamating process results in a compact insulation material after a few days.

With paper-insulated cables, the insulation of the flexible joint has a similar structure as the cable itself, comprising a conductor screen, the electric insulation, and an insulation screen. The screens are made from carbon-black paper tapes. The paper tapes are pre-impregnated and are applied either manually or with a semi-automatic lapping machine (Fig. 4.2). The joints should be manufactured in humidity-controlled and temperature-controlled rooms.

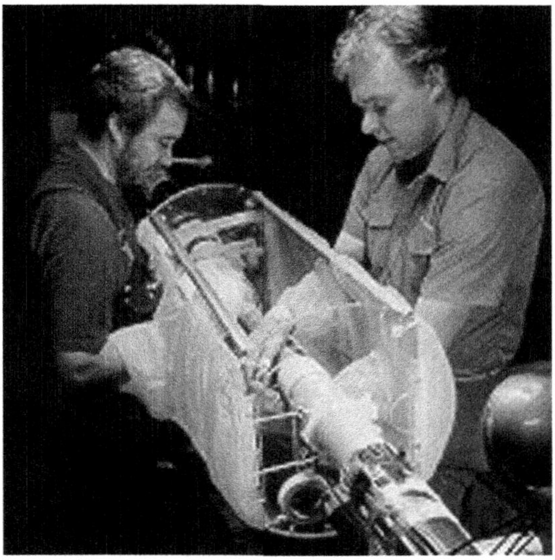

**Fig. 4.1** Application of XLPE joint insulation under clean conditions (Courtesy of ABB, Sweden)

**Fig. 4.2** Application of paper insulation for a flexible joint using a semi-automatic lapping machine (Courtesy of ABB, Sweden)

The factory joint also includes a lead sheath over the jointed insulation. The lead sheath is applied as a wide lead tube, which is slid over and parked on one side of the factory joint before the conductor jointing. After the insulation is finished, the lead tube is pushed over the factory joint, swaged down to a tight fit over the joint insulation, trimmed, and soldered to the cable's lead sheath. In case of mass-impregnated cables, the insulation inside the lead sheath can now be treated again to achieve fully saturated paper-oil insulation. Finally, a protective polymeric shrink tube is applied over the lead or other outer layers. Now, the factory joint constitutes an integral part of the power cable core and is ready for further steps in the production line such as armoring. The slight oversize of the factory joint does not impose any obstacle for the further production.

### *4.1.2 Offshore Installation Joints*

The notion "installation joint" or "field joint" describes a joint of the complete submarine power cable including conductor, insulation system, armoring and all intermediate layers.

Installation joints are manufactured on-board of a sea-going vessel, or in the beach area. Depending on the cable and joint design, the manufacturing of offshore installation joints takes between one and ten days after both cable ends are laid up in the jointing shack on-board the vessel. No matter if this is the laying vessel or a separate jointing barge, the joint should be designed to keep the offshore jointing time as short as possible. During the jointing, at least one of the cables is hanging down from the vessel over a laying wheel or a laying chute.[2] In heavy weather this imposes a risk for the jointing crew and the cable. If the required jointing time

[2]One of the cables to be jointed may still be on the turntable on the cable laying vessel.

is short, there are better chances to find a sufficiently long weather window with suitable sea-state. Once the jointing process is started it can be interrupted only by cutting the cable.

#### 4.1.2.1 Flexible Installation Joints

Flexible joints can be used with advantage when a long cable route requires the offshore jointing of subsequent delivery lengths. After the first laying campaign, the vessel would fetch the next cable length at the manufacturer's premises or a storage port. It returns to the cable route and pulls up the end of the first cable over the laying wheel to a jointing shack onboard the vessel. There, the second cable length, which is still onboard the vessel, is jointed to the first cable by means of a flexible installation joint.

The basis of a flexible repair joint is the same procedure as for the factory joint, including the jointing of conductor, insulation, and lead sheath as described for the factory joint. The joint may be slightly thicker than the corresponding cable diameter. Finally, and this is the special attribute of submarine power cable joints, the armoring over the installation joint must be arranged with high tensional strength, yet flexible.

For flexible joints a wire armoring covering the joint section is necessary. The gap between the armoring of the two cable ends can be closed with pre-spiralled wires. These wires are welded to the wires on one side of the joint, twined around the joint, and then connected to the armoring wires on the other side of the joint. A pretension of the joint armoring wires is necessary to maintain the tensional forces in the armoring. The armoring wires can be welded to each other wire by wire, or in various welding patterns to achieve the best possible tensional strength between the cable armoring and the joint armoring. The wires of the joint and the cables can be connected also by screw connectors [3] or by what is known as turnbuckle, a left/right-threaded sleeve to tighten the connection between armoring wires [4]. The use of ring-shaped welding sleeves to connect the wires from the joint to the cable armoring wires provides a good distribution of tensional forces. A welding sleeve makes it also possible to connect cables with different numbers, diameters or shapes of armoring wires. A Swedish manufacturer uses a patented [5] semiautomatic portable armoring machine to provide the joint armoring.

A picture or illustration of a flexible repair joint would not be very informative as the joint has the same appearance as the adjacent cables.

The completed joint can now be transferred through the ordinary laying equipment of the vessel and passes over the stern laying wheel down into the water. The jointed new cable length is following and the laying operation can be continued. Depending on the cable design, the manufacturing of a flexible joint requires a jointing shack of 4 up to 18 m length on-board, located between the storage turntable and the stern laying wheel.

Flexible cable joints can be made for single-core mass-impregnated cables of any rated voltage. For polymeric single-core cables, flexible installation joints have been used for voltages up to 245 kV. Flexible joints have also been used for three-phase cables up to 150 kV [6].

Cable factories without their own sea harbour can produce submarine cables only in short drum lengths fit for road transport to the next port. In these cases, flexible joints are used to connect a number of drum lengths to the desired submarine cable delivery length, which is an inconvenient and tedious method [4, 6].

#### 4.1.2.2 Rigid Joints

Rigid joints (or "stiff joints") are very different to flexible joints. The name denotes that the joint has a rigid outer casing, most often in the shape of a steel tube (Fig. 4.3). The steel tube serves as a connecting point for the cable armoring wires of each cable end, and as an outer protection to the cable joint inside.

The jointing of the electric systems inside the outer casing can be accomplished in different ways. The cable core ends may be connected with the flexible jointing method as described above. The outer steel casing provides additional mechanical strength and tensional force. In another method, the steel casing allows to use pre-moulded joint sleeves for the electrical parts of the cable. The pre-moulded joint design employs an elastomeric sleeve containing both semi-conducting and insulating layers. This sleeve is bridging the gap between the cable insulation on each side (cf. Fig. 4.4).

The concept of pre-moulded or pre-fabricated joints has become the method of choice for polymeric land cables because of many advantages:

- short assembly time
- accommodates all types of conductor connections such as welding, compression sleeves and screw connectors according to client specifications
- can be pre-tested in factory

The steel casing makes it possible to use pre-fab joints also in submarine applications. Sometimes, an inner casing made of copper or brass is used which encloses

**Fig. 4.3** Rigid submarine installation joint in a steel casing for a 1C cable during an on-shore bending test

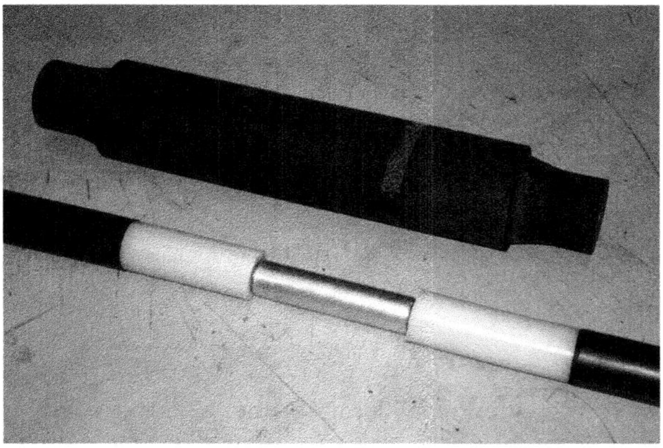

**Fig. 4.4** Pre-moulded polymeric cable joint. The elastomeric joint sleeve on *top*. Conductor connected and insulation prepared for application of insulation sleeve on *bottom*

the pre-fab joint and is soldered to the lead sheath for complete water tightness. Before jointing the conductors of the two cable ends, the pre-fab sleeve is expanded radially and "parked" on an auxiliary support tube on the nearby power cable end. After conductor jointing and preparing the insulation according to the instructions of the supplier, the "parking" sleeve is slid over the insulation gap. The support tube is removed, and the joint sleeve collapses over the insulation gap in a pre-determined position. In this final position, the sleeve keeps a radial pressure onto the underlying insulation surfaces. The radial pressure between the interface surfaces is of critical importance for the dielectric strength of the joint over the total lifetime. In some designs, external spring-loaded elements help keeping that radial pressure up.

In contrast to the taped or tape-moulded joint, the pre-fab joint sleeve can be pre-tested at the manufacturer's premises. The high-voltage factory test, often with partial discharge recording, can detect possible flaws, voids, or particles in the insulation material and confirms the intrinsic dielectric strength of the tested joint sleeve. Naturally, this test cannot unveil a poor installation job.

The core joint is equipped with its own water barrier system (very often a lead sheath). If the pre-fab joint sleeve is very bulky, a copper or brass casing can be used to encapsulate the joint. The copper casing can easily be soldered to the cable lead sheath in order to provide a completely watertight cover for the joint.

For some medium voltage cables, a polymeric water protection is considered adequate and the joint can be covered with a shrink tube.

The jointed core(s) are now encased into the outer metallic casing. The casing has a central cylindrical portion and two conical ends. The partition line of the steel casing may run in parallel with the cable axis connecting two half-shells, or may run around the ensemble connecting two funnel-like parts. The outer casing can also be assembled from more than two parts. The different parts of the steel casing are connected to each other by welding or nuts and bolts. The wire armoring on each side of the joint casing is connected mechanically to the casing by means of clamping flanges or welding (cf. Fig. 4.5).

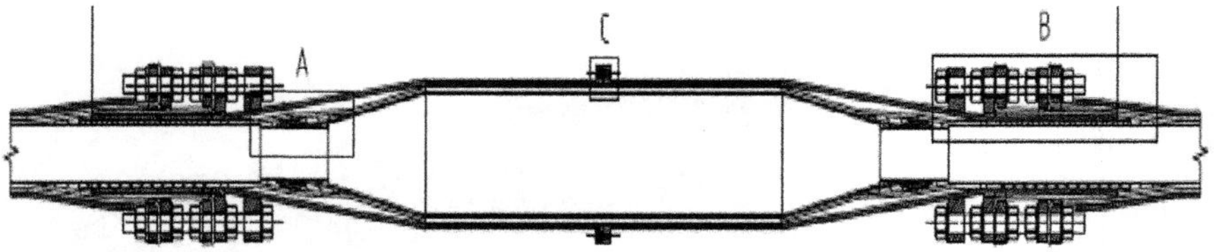


**Fig. 4.5** Copper and steel casing for a pre-moulded joint

In Fig. 4.5, the soldering of the copper casing to the cable lead sheath can be seen at "A", and the partition of the steel casing at "C". The casing constitutes a strength member connecting firmly the armoring layers of each cable side (at "B"). For less demanding applications such as shallow waters, a casing made from polymeric materials may be used. Special attention must be paid to the transition between the layers of the cable and the corresponding layers of the joint as the manufacturing, transport, installation, and service can impose high mechanical stresses on all parts of the joint.

In order to avoid sharp cable bends at the transition between the stiff joint casing and the flexible armoring of the submarine cable, bend restrictors in the shape of conical rubber sleeves enclose the cable at the exit of the steel casing (cf. Fig. 4.3).

Rigid joints cannot be transported through the ordinary cable gantries or deployed with the ordinary vessel laying equipment because of their stiffness and increased diameter. Complicated crane arrangements are necessary for the deployment of rigid joints.

All these facts seem to make the rigid joint the inferior choice of installation joints for submarine power cables. Indeed, the flexible joint as outlined in the previous chapter has the advantage of simple design and easy installation. But the advantages of pre-fab joint sleeves and the good mechanical protection provided by the steel casing may outweigh the use of extra components. At the end, assembly time on board the vessel is very precious.

While rigid joints have been used for submarine LPOF cables, they are not being used for today's MI d.c. cables. As there are no pre-fab MI cable joints available, the joint insulation must be made by on-site lapping anyhow. Rigid joints can therefore not provide the advantage of quick assembly and factory pre-testing for mass-impregnated cables.

Rigid installation joints are used for almost all 3C cables (cf. Fig. 4.6). Inside the joint case, the three cable cores are split and jointed individually. Each cable core joint is encapsulated in its own water-proof inner casing, which can be made from pre-formed copper tubing to be soldered to the lead sheath of the cable. The functional separation of inner casing (water protection) and outer steel casing (tensional forces, armoring) allows for more possibilities in the design process. The outer casing also can accommodate a splice box for optical cables, which might be incorporated into the power cable.

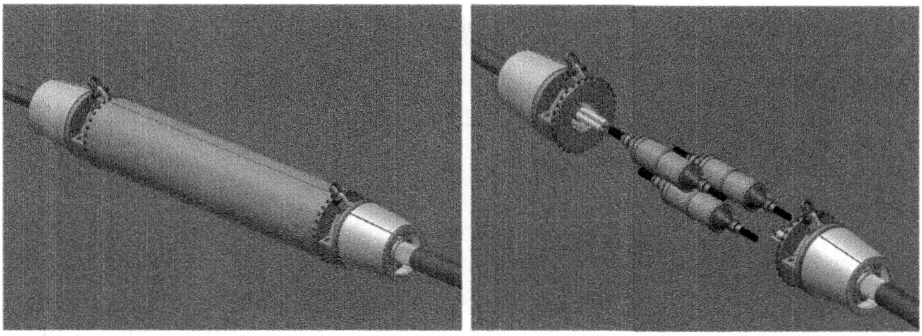

**Fig. 4.6** Rigid joint casing with three pre-moulded joints inside (Courtesy ABB, Sweden)

### *4.1.3 Miscellaneous Joint Designs*

Fluid-filled submarine power cables require joints with a continuous fluid duct in the hollow conductor. A tubular sleeve is often included into the duct to provide support before welding. The cable cores are frozen down by liquid nitrogen to prevent oil outflow.

Under certain circumstances, the free flow of impregnating fluid through the joint is not wanted, e.g. in steep passages of the submarine cable. In these cases, stop joints can be devised, which have a solid conductor weld, closing both the central oil duct and the inter-wire spaces. Conical epoxy resin spacers inserted into the insulation wall stop the oil passage in the insulation. These stop joints can also prevent unwanted movements of copper particles with the oil [4].

Many transition joints have been designed to connect submarine cables of different kinds. Submarine cable links sometimes comprise different conductor sizes to match different thermal conditions along the route. Small size differences can be jointed with flexible joints. For large size differences, the use of a rigid joint is preferred because it provides a protection against mechanical stress, which can be detrimental to the sensitive transition of the highly stressed conductor screen.

Some submarine 150 kV HVDC projects with extruded cables have copper conductors in the submarine portion and aluminium conductors in the land section (Estlink, NordE.ON). The transition between the different cables was accomplished by purpose-made pre-fab joints with two-metal screw connectors. The transition joint connected a 1000 $mm^2$ Cu submarine cable conductor to a 2000 $mm^2$ Al land cable conductor.

In some submarine HVDC cable links the near-shore cable is a fluid-filled cable, while the submarine cable is a mass-impregnated cable [7]. Also here transition joints can be constructed to connect cables with different insulation systems and different conductor design. Transition joints between paper-insulated cables and extruded cables are known from on-shore applications but no case of submarine use is known.

**Table 4.1** Application and properties of various joint concepts

| | Flexible joint | Rigid joints, tape-insulated | Rigid joints, pre-fab |
|---|---|---|---|
| 1C mass-impregnated | Yes, all voltages | Yes, but no advantages | No |
| 1C paper-insulated | Possible | Possible | No |
| 1C extruded cables | Up to 145 kV. In few cases 245 kV. | Common up to 110 kV | Yes |
| 3C paper-insulated | Possible | Yes, but no advantages | No |
| 3C extruded cables | Possible | Yes, but no advantages | Yes |
| Conductor joint | Welded only | Welded only | Welded, compression or screw |
| Need for special deployment arrangement on cable vessel | No | Yes | Yes |
| Space for optical joint box | 1C cable: No | 3C cable: Yes | 3C cable: Yes |

Trifurcation joints connecting a three-core cable to three single-core cables are known for land applications but no submarine use is known. Probably trifurcation joints have been used as beach joints to connect 3C submarine cables to 1C land cables.

A bifurcation joint has been manufactured for the NorNed project; it connected a two-core HVDC cable in the southern part of the cable route to two single-core cables in the northern part.

The large variety of submarine cable joints may cause some confusion. Table 4.1 lists those joint concepts described here and their most important properties.

### *4.1.4 Beach Joints*

Sometimes, the submarine cable can be connected to an overhead line or a substation directly at the landing point, but in many cases the cable route continues onshore. A beach joint between the submarine and the land cable can be necessary for different reasons:

– The cable termination is too far from the shoreline to pull in the submarine cable all the way.[3]
– Thermal conditions require a larger conductor size onshore than offshore. It would be imprudent to dimension a long submarine cable for the thermal needs of a short beach section.

[3]The longest pull hitherto has been made when the Swedish end of the Baltic Cable was pulled from the sea more than 4 km onshore to the cable-to-air transition yard.

- In many projects the submarine cable is 3C while the land cable system consists of three single-core cables. The transition joint is usually erected in the shoreline

The design and installation of beach joints may require special attention. The location for the beach joint should be selected carefully if there is a choice. A dry jointing house or tent offering protection from wind and airborne sand, salt, and droplets, is necessary for the assembly of a joint. The safest place is on dry land, sufficiently far away from the splash zone. The higher the location over the water table, the easier it is to keep the joint pit dry. A location beside a road would facilitate transport of equipment and minimize impact on the environment.

The manufacturing of the beach joint in the splash zone generates further difficulties. The jointing pit must be excavated into the near-shore seafloor, secured by a cofferdam and pumped dry. The cable ends must be guided through the cofferdam walls. Equipment and crew must be given a dry clean environment. Alternatively, the joint can be erected on a lofty platform. In either case, the work preparations are even worse when tidal current threaten the stability of the arrangements twice a day.

Figure 4.7 illustrates the beach joint area for a double HVDC extruded cable circuit. Two pairs of HVDC cables are pulled in from offshore through pipes into land. The protection pipes are terminated in beach holding devices that are bolted into the concrete floor of the joint pit.

**Fig. 4.7** Beach joints of two pairs of extruded HVDC cables

## 4.2 Cable Terminations

The submarine cable, when landed onshore, is normally connected (jointed) to an underground cable close to the beach. The underground cable continues to a substation, where it is terminated with a standard onshore cable termination. The substation can be a few meters or many kilometres from shore.

The onshore cable termination is not specifically designed for submarine cables but for underground cables.

### *4.2.1 On-Shore a.c. Cable Terminations*

Onshore a.c cable terminations are equal for submarine and underground cables in most respects. Standard terminations for a.c. submarine cables are available from a row of manufacturers:

- *Open-air terminations* to connect the cables to overhead lines or the busbars of an air-insulated substation. The insulator may be of porcelain or polymeric. The creepage length over the insulator sheds is specified between 25 and over 40 mm/kV depending on the expected salt and pollution load. Expected salt storms close to shore also require special attention to the corrosion protection of the metal work of the termination. It is recommended to replace standard aluminium grades to highly corrosion resistance grades. Also, specified wind loads for near-shore terminations might be more stringent.
- *GIS terminations* to connect the cables to gas-insulated switchgear. This type of cable terminations is well-known from land cable applications up to the highest voltages. GIS terminations have standardised sizes in order to make them compatible to switchgear from other manufacturers. GIS terminations are based on a stress relief cone on the extremity of the cable insulation. The stress cone is inserted into a conical receptacle inside the GIS compartment. In some GIS terminations, the stress cone matches the inside of the conical receptacle perfectly (plug-in type), and these terminations are completely oil-free. Other designs contain a small amount of dielectric fluid to fill the gap between the stress cone and the inside of the conical receptacle.
- *Transformer terminations* are generally identical or very similar to GIS terminations. However, for submarine cables they are rarely used, as substation planners want to have circuit-breakers between the submarine cables and the transformer.

### *4.2.2 On-Shore d.c. Cable Terminations*

So far, only cable-to-air terminations have been devised for submarine d.c. cables. The stress control in d.c. termination must rely at least partly on resistive elements,

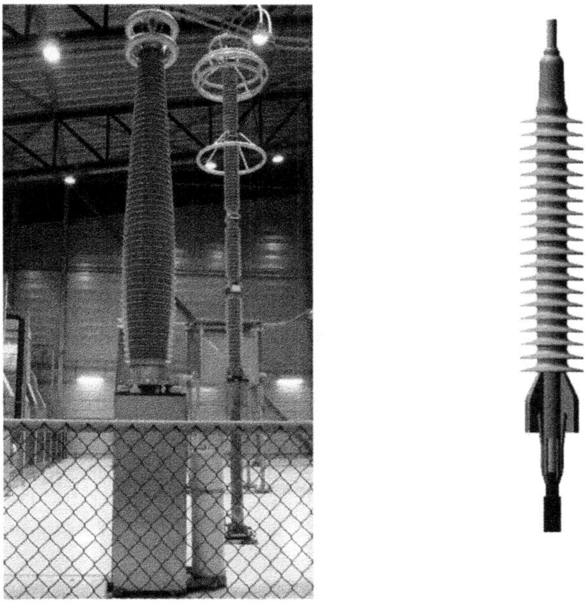

**Fig. 4.8** Indoor HVDC cable terminations for mass-impregnated cable (*left*) and extruded cable (*right*). Small oil reservoir cylinder is attached to the stand of the left termination

as pure capacitive elements would be not suitable with d.c. The design is adapted to the type of submarine d.c. cable.

The terminations used for extruded submarine HVDC cables up to 150 kV are identical to on-shore cable terminations for the same cable technology. They have a polymeric insulator and are completely oil-free. Terminations for extruded HVDC cables are erected indoors.[4] The termination shown in Fig. 4.8 has a field-grading element that is made of a resistive material with non-linear resistivity. This material becomes more conductive when subject to higher electric stress. Owing to this property the field grading element has the ability to relieve highly stressed parts inside the termination.

LPOF cable terminations for d.c. use are designed very similar to their sisters for a.c. use. However, the stress cone and the external insulator might have a different design to control the d.c. stress. The stress cone is made from impregnated paper, which is either pre-formed in the factory or shaped on-site during the erection of the termination. The termination also constitutes the oil-feeding entry point into the cable.

The terminations for mass-impregnated HVDC cables are similar to the LPOF cable terminations except the oil-feeding system. While the LPOF cable termination must take care of the thermal "breathing" of the entire cable (or at least half of the length of it), in the mass-impregnated system only the expansion of the small oil

[4]It is yet not known how the cable terminations for the Trans Bay HVDC project (under construction) will be arranged.

volume inside the termination must be taken care of. For this purpose, a relatively small oil expansion vessel of a few hundred litres is enough and can be mounted on the termination stand without external pipework (cf. Fig. 4.8).

### *4.2.3 Offshore Cable Terminations*

Submarine power cables connected to offshore installations such as oil and gas production platforms, or OWP must be terminated in a harsh environment. The adverse climate and restricted space allow the use of open-air terminations only for moderate voltage levels. The cables are often terminated directly into encapsulated switchgear by means of GIS terminations, polymeric plug-in connectors, or transformer terminations.[5] Again, these components are standard components from land cable applications. However, they must be corrosion proof and comply with the product and safety standards that rule onboard the platform. These standards may be more stringent than equivalent standards onshore. Also, the erection and installation of the terminations may be subject to much stricter rules compared to those that apply for onshore work.

## 4.3 Other Accessories

A number of accessories are used for the structural integration and safe fixation of submarine power cables. The following lines give but a rough overview on the large variety of products.

### *4.3.1 J-Tubes*

It is industrial practice to guide power cables up to stationary platforms through J-tubes, named for their J-like shape. The bow of the J is down on the seafloor and the upper end of the J is beneath or above the lowest platform deck. The lower opening is called bellmouth for its shape and is normally directed outwards from the platform legs. The bellmouth can be underneath or slightly above the seafloor. During the installation, the cable is pulled through the bellmouth up to the platform by means of a pulling wire. In order to guarantee a smooth installation, the bow radius should be noticeably larger than the minimum bending radius of the cable, and the tube diameter at least 2.5 times the cable diameter. It is cheaper to add some material to the construction than to get stuck with a cable and an expensive cable vessel scheme during the installation. Normally, there is one power cable in each J-tube, but numbers of two and four have been used.

---

[5] The terminations of extruded HVDC cables on offshore platforms are usually placed inside the valve cubicles.

While most J-tubes are left open at the bottom, some are plugged around the cable in order to keep anti-corrosion fluid inside.

The thermal conditions of the power cable inside the tube should be given special attention. In the water-filled J-tube, convection contributes to the heat flow between the cable and the tube wall. Convection is depending on the size of the annular gap and is difficult to grasp mathematically. The conditions get worse in the upper air-filled part of the J-tube where the air can get trapped inside the tube. Sufficiently large openings on top and bottom of the air column create a chimney effect and improve the situation considerably. However, openings close to the splash zone are not liked by corrosion specialists.

Some central platforms in OWP's have giant J-tubes that accommodate a large number of incoming cables from the individual chains of turbines. It can get crowded and warm inside.

### *4.3.2 Hang-Off*

The gravity weight of vertically suspended cables on stationary or floating platforms is carried by hang-offs. A hang-off is a sophisticated connection flange between the cable armoring and the platform structure (cf. Fig. 4.9). The flange contains a clamping device for the armoring wires of the power cable to carry the mechanical load. The cable core with lead or copper sheath and protective plastic sheath is extending through the hang-off and continues upwards towards the cable termination.

**Fig. 4.9** Hang-off for a three-phase cable connecting an offshore wind turbine

### 4.3.3 Bending Protection

As any flexible product, a submarine power cable is sensible for overbend and fatigue where there is a discontinuity in bending stiffness. This situation can be found at a cable entrance into a rigid joint enclosure cable, entrances to fixed structures such as hang-off, or cable glands into floating structures. Overbends may occur, or repeated bending may cause severe fatigue in the cable construction. Bending stiffeners are elastomer sleeves that engulf the power cable close to the entrance into a rigid structure. The conical shape of the bending stiffener provides a gradual increase of bending stiffness and defines a gentle bending curve for the flexible cable (Fig. 4.10). The bending stiffener must be designed for the specific case.

A bending restrictor consists of a number of interlocked polymeric or metal shells around the cable allowing for a certain bending angle for each interlocked shell. The bend restrictor defines a minimum bend radius regardless of the cable load. As it has a stepwise bending stiffness adding to the cable bending stiffness, there may be a discontinuity of bending stiffness anyhow. With improper design, the sharp-bend problem will just be re-located to the end of the bend restrictor.

**Fig. 4.10** Bending stiffener (*right hand*) together with a diverless subsea cable connector (*left side*). (Courtesy of Trelleborg, UK)

### 4.3.4 Holding Devices

Various clamping devices can be used to secure submarine power cables in beach areas, along steep underwater slopes, in areas of strong currents, and elsewhere.

## References

1. HVDC Development options – Cable Capacity. 3rd Supporting Document to the Investment Proposal for the HVDC Inter-Island Link Upgrade Project, Transpower New Zealand Ltd, 2005. http://www.electricitycommission.govt.nz/pdfs/opdev/transmis/gup/Vol3/HVDC-Supporting-docs/3-Cable-capacity.pdf

2. Foxall R G et al. (1984). Design, Manufacture and Installation of a 525 kV Alternating Current Submarine Cable Link from Mainland Canada to Vancouver Island. Paper 21-04, Cigré Session 1984, Paris, France.
3. Pirelli Cable Review, June 1971.
4. Elgh L et al. (1974). The 420 kV Submarine Cable Connection Denmark – Sweden. Cigré Paper 21-02, Cigré Session 1974.
5. European Patent EP1320915. Published 2003.
6. Galloway S J et al. (1990). 150 kV Java-Madure Submarine Cable System Interconnection, Power Engineering Journal, January 1990, p. 7–15.
7. Giorgi G et al. (2002). The Italy – Greece HVDC Link, Cigré paper 14-116. Cigré Session 2002, Paris, France.

# Chapter 5
# Manufacturing and Testing

## Contents



## 5.1 Manufacturing

The manufacturing of land cables is a straight-forward consecutive row of adding layer by layer to the conductor. The short production length (typically a drum length) allows for moving the cable between the different production steps on drums until the final layer has been applied and the cable drums can be put to the final acceptance test.

Submarine power cables on the contrary are often shipped in extreme lengths only limited by the capacity of the cable-laying vessel, which can be several thousand tons of cable. To avoid a large amount of joints, all production steps must be performed on lengths as long as possible. The maximum continuous production length of each production step depends on the cable size and type. Most often


T. Worzyk, *Submarine Power Cables*, Power Systems,
DOI 10.1007/978-3-642-01270-9_5, © Springer-Verlag Berlin Heidelberg 2009

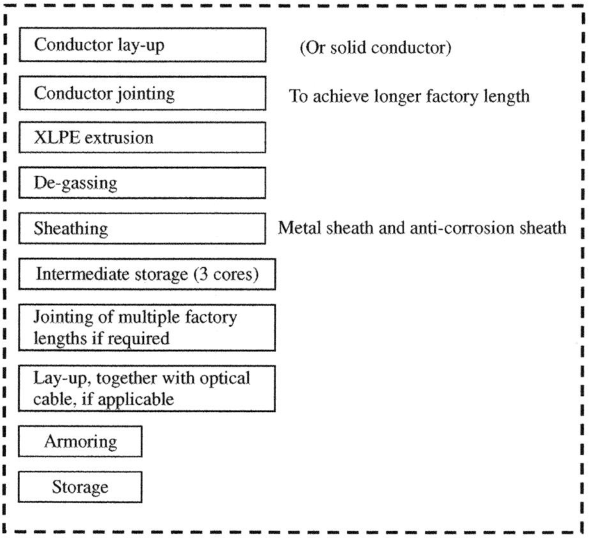


**Fig. 5.1** Flow chart for the production of three-core submarine cables with extruded insulation

the production lengths of the different steps cannot be harmonized completely. This results in an intricate planning of the use of production assets. Only a well-orchestrated factory can exploit the advantage of long shipping length.

Figure 5.1 is a flow chart of the manufacturing of a three-core extruded submarine cable.

The core of a cable consists of a conductor, the electric insulation system, and protecting sheathes such as lead and PE sheath. The manufacturing of submarine cable cores is almost identical to the manufacturing of land cable cores and is described here but briefly. More detailed descriptions of cable core manufacturing can be found in [1, 2]. In this chapter the focus is on those manufacturing steps that are characteristic for submarine power cables.

### *5.1.1 The Conductor*

All cable manufacturing starts with the conductor. Conductor making methods are identical to those for underground cables. However, the conductor made from pre-shaped profiled Cu or Al wires is used almost exclusively for submarine HVDC cables where a very compact conductor has a larger advantage than in an a.c. cable. For mass-impregnated cables it is essential that the wires lay closely together not leaving any space for extra impregnation compound [1]. Wire design and manufacturing process must be matched carefully to avoid gaps between the wires, or stepping, which is a misalignment of the profiled wires leading to a saw-teeth like surface on the conductor. A well-adjusted conductor stranding line makes a perfect smooth conductor. Conductor sizes up to around 3000 $mm^2$ have been achieved.

Some factories can produce conductors in continuous lengths up to 20–30 km but most factories produce the conductor on cable drums in lengths of only a few kilometres. Conductors can be jointed before entering the next production step by a multitude of jointing technologies:

- Solid welding in one piece through the cross section
- Wire-by-wire welding
- Soldering or brazing techniques
- Flush compression sleeve.

Conductor joints for fluid-filled cables (not mass-impregnated) require an oil passage in the centre. All conductor joints must be performed very carefully to provide sufficient mechanical strength and a smooth flush surface. In particular, "cold" soldering must be avoided as the wire ends might snap apart during later bending of the cable and perforate the conductor screen from the inside. The conductor joint must also meet the stringent tensile requirement imposed on submarine cables for larger depths. The area on both sides of the welding is often softer than the weld or the conductor itself on account of the annealing effect of the heat.

A frequent requirement for submarine power cables is longitudinal water tightness in the conductor. This can be achieved by the introduction of various water-blocking agents such as swelling powder, swelling yarns, hydrophobic compounds or gels. They are inserted between the wire layers.

The conductor resistance should be checked on a regular basis according to the quality assessment system of the manufacturer.

The next step in the manufacturing process is the application of the insulation. In modern submarine power cables, only extruded XLPE or lapped paper insulation is being used. These two processes are completely different in material, equipment and production properties.

### *5.1.2 XLPE Cables*

XLPE extrusion is a continuous process. The raw material is a highly sophisticated compound of base polyethylene resin with well-defined distribution of molecular weight, antioxidants and cross-linking agents. The compound is supplied as granules and comes in different purity classes. In general, a higher voltage class requires a higher purity. The properties of the semiconducting compound forming the conductor screen and the insulation screen are tailor-made to fit the insulation compound. There are smooth and ultra-smooth varieties regarding the smoothness of the highly stressed inner semi-conductive surface.

Virtually all modern m.v. or h.v. cable extrusion lines employ a triple-extrusion concept. Three extrusion screws supply the constituent materials to a common extrusion head, which extrudes all three layers concentrically in the same process. Catenary or vertical extrusion lines are used to manufacture the insulation system. While an extrusion length of 2000 m is sufficient to produce underground cables in drum lengths, the long delivery length of submarine cables call for much longer uninterrupted extrusion length in order to reduce the number of joints. Lengths of over 20 km can be extruded in a single run. The long extrusion time requires a faultless supply of resin granules under super-clean conditions. In the best cable extrusion lines there are closed systems for the handling of the granules to avoid any contact with factory air or human hands.

Any interruption in the process causes a cut of the cable rendering shorter cable lengths than anticipated.

Extrusion lengths, whether limited by extruder limitations or by unintended cable cuts, can be jointed together using flexible factory joints to achieve long delivery lengths. Flexible factory joints are provided with a conductor connection of one of the types outlined above. The conductor joint must be flush with the cable conductor. After the conductor connection is made the insulation of the adjacent cable ends are tapered. The gap is filled with lapped XLPE tapes and cured under heat and pressure. The added insulation material melts together with the surface of the tapered cable insulation. Figure 5.2 shows such a flexible factory joint. Some factories produce injection-moulded factory joints. The flexibility of the joint is crucial for the subsequent path of the cable core through the cable gantries of the factory.

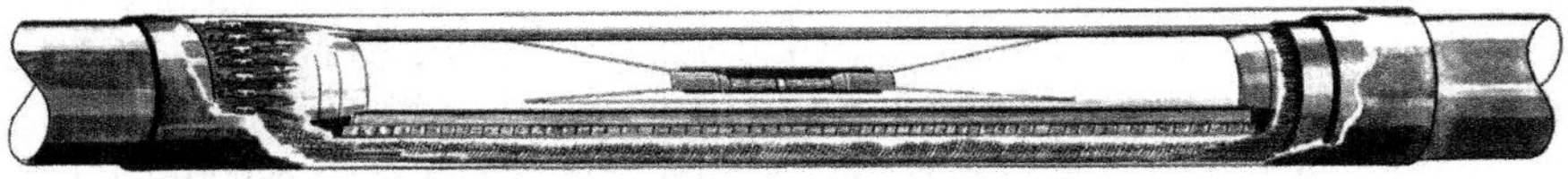

**Fig. 5.2** Factory joint of an extruded cable core (Courtesy of ABB, Sweden)

Gaseous byproducts that are dissolved in the XLPE matrix must be removed by degassing. If not degassed properly, the cables would emit these gases during operation later on resulting in unwanted gas volumes in the conductor or under the lead sheath. The degassing of long lengths of submarine cable cores requires large heated and ventilated vessels.

The insulation material must be protected against humidity already in the factory. Condensing water can be found on resin pellets cold from the transport. At cable cutting surfaces humidity may enter the conductor and contaminate the insulation from inside. The manufacturer's QA system defines when and where during storage time the cable must be equipped with suitable end caps.

### *5.1.3 Paper-Insulated Cables*

The paper insulation is applied on dedicated paper lapping lines consisting of a row of lapping heads. In each lapping head, typically 12–16 reels with paper tapes revolve around the cable conductor as it passes through.

The paper tapes are applied according to an intricate pattern. Each single paper tape is wrapped around the cable in an open helix (it does not overlap itself in

**Fig. 5.3** Lapping of carbon-black paper and insulation paper onto a copper conductor (Courtesy of ABB, Sweden)

subsequent turns). There must be a small gap (1–4 mm) between consecutive turns. This gap is called butt gap. When the cable is bent, the butt gaps in the inner curve accommodate the relative movement of the individual paper tapes.

The paper tension is controlled for each paper tape by brake devices on each reel. The lapping heads rotate in alternate directions to produce a torque-balanced insulation. Figure 5.3 shows the cable core in a lapping head. A layer of semi-conducting carbon-black paper is lapped onto the conductor before the insulating paper is added. High-performance power cables require the use of pre-dried cable paper and a lapping line in a controlled dry atmosphere.

Many cable types use staggered paper thicknesses throughout the insulation, each layer having a well-defined paper tension. The paper thickness ranges between 40 μm and 180 μm.[1] The insulation can have as many as 270 individual paper tapes to be put on the conductor in a single run.

The paper lapping process can be stopped and started at will. Reloading of the paper reels consumes a considerable part of the production time, and it is an appreciable logistic effort to supply sufficient paper reels of the correct type in time. After completion of the insulation including the screen the cable core is guided to the receiving turntable. This can be an intermediate storage turntable, or the cable is being stored directly into the rotating impregnation vessel. In either case the cable must be kept under very dry air.

It is, drawing on sufficient experience, and obeying strict quality standards, possible to produce a firm regular insulation without wrinkles or creases over very long lengths. The process is virtually unlimited as the next conductor length can be welded to the previous one upstream of the lapping line. Only the hold capacity of the receiving turntable limits the uninterrupted lapped cable core length.

While paper in equilibrium with the open air holds a water content of 6–12% the accepted level for paper insulation is only within the range of 1–2%. The

[1]Paper suppliers prefer to specify the paper "thickness" in grams per square metres (g/m$^2$).

lapped insulation must therefore be dried before impregnation. Today, major cable makers have very large drying vessels to process up to 50 km of cable core in a batch.

Once the lapped cable is in the impregnation vessel the lid is closed and the cable inside is subjected to a heat/vacuum treatment as the final drying step.

So far, the process is identical for mass-impregnated and for LPOF cables. The subsequent impregnation and sheathing processes differ between the two insulation concepts owing to the very different viscosity of the impregnant.

The impregnation compound for mass-impregnated cables (today almost invariably used for HVDC cables) becomes rather liquid at the impregnation temperature of 120°C (Fig. 5.4). The compound is heated and de-gassed in a dedicated oil treatment plant in the cable factory before it is released into the closed impregnation vessel. The vessel is filled completely, and under pressure. Owing to the lower viscosity of the hot compound the impregnation is finished after a few days. However, the cable cannot be processed yet. The cooling of mass-impregnated d.c. cables must be controlled carefully as the contracting compound in the insulation must be completed by new compound from the free vessel volume. As the temperature decreases, the compound viscosity increases, and makes the replenishment more time-consuming. If the cable is taken out to the lead sheathing too early the subsequent cooling would render unwanted voids in the insulation. The cooling time for mass-impregnated cables can sum up to several weeks. Once the cooling criteria have been met the vessel lid can be lifted and the cable can carefully be pulled to the lead sheathing machine. The short-time contact to the free air does not harm the cable insulation since the high viscosity of the compound prevents the diffusion of soluble nitrogen or humidity into the insulation.

LPOF (or SCFF) cables will be impregnated with very low viscosity mineral oil synthetic impregnation fluids. The fluid is degassed and prepared before it is pumped into the vessel to soak the dried cable core. Thanks to the low viscosity the impreg-

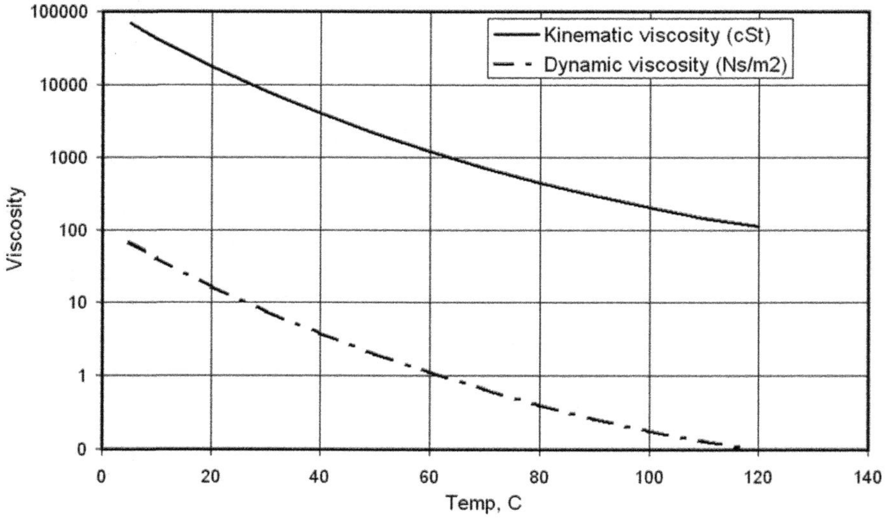


**Fig. 5.4** Viscosity vs. temperature of a HVDC cable impregnation mass [3]

nation and cooling process is much faster than for mass-impregnated cables. The impregnated cable core of oil-filled paper-insulated cables cannot be transported from the impregnation vessel to the lead-sheathing machine through open air as the low-viscosity oil would eagerly adsorb air humidity. Instead, an oil-filled tube carries the insulated conductor from under the oil level of the impregnation vessel over the rim and into the lead extruder. The tube is continuously flushed with degassed impregnation oil.

### *5.1.4 Sheathing*

Most submarine power cables have a radial water barrier in form of a metallic sheath.[2] Shorter medium-voltage cables often have an aluminium laminate sheath consisting of aluminium foil coated with a thermoplastic layer. The aluminium laminate is folded longitudinally around the cable core and glued together, sometimes with a Z-fold. The subsequent plastic sheathing provides mechanical support to the laminate.

The lead sheath for submarine power cables is applied in uniform thickness between 2 and 5 mm with a lead press, or an extruder. In the press, also known as ram press, molten lead is being filled into a chamber and cooled down to the correct temperature under the melting point. Then the lead is pressed by hydraulic force through a die forming a seamless tube around the cable, which travels slowly through the press. When the lead chamber is emptied, the process stops and the chamber is being refilled with molten lead. The intermittent process causes change marks in the cable sheath each time the press stops [1]. Another lead sheathing machine is the Hansson-Robertson continuous lead extruder, which was commercialised in 1949. Basically it is a screw extruder fed with lead from a melting pot. Today Hansson-Robertson extruders can deliver 50 km and more of uninterrupted high-quality lead sheath. Continuous extruders have replaced the ram-press as the most common and economical means of producing lead sheathed power cables. Different lead sheathing technologies require different lead alloys for best performance and process stability. The development and properties of different lead alloys in connection with ram press and screw extruder is described in [4].

Long term stability, creep, and extrusion properties can be improved substantially by using lead alloys with alloy elements such as antimony, tin, copper, calcium, cadmium, tellurium, and others. The standard EN 50307 lists a number of lead alloys for cable use (cf. table of lead alloys in Chap. 11).

The lead sheath is vulnerable for mechanical damages and should be protected by additional layers as soon as possible. The use of a continuous lead extruder allows for the extrusion of a rugged plastic oversheath in a tandem process directly after the lead extruder. When the cable has been lead sheathed in a stop-and-go ram press, it must be stored on an intermediate turntable before the plastic oversheath can be applied in the next operation.

[2] In older books on the subject, the word "sheathing" is used for "armouring".

If the cable should be stored temporarily in a bare lead sheath it can be useful to brush the lead sheath with a bituminous solution to prevent the lead sheath from sticking together.

Although it is possible to produce aluminium sheaths by extrusion or strip welding this material is, due to its poor corrosion resistance, hardly ever used for submarine cables. In some cases the sheath is made from copper. A copper strip is folded longitudinally around the cable, and the edges are trimmed and welded continuously. Then the formed copper tube is corrugated to increase the flexibility. So far this process has been used only for shorter cable lengths in the range of a kilometre. As a perfect welding seam is fundamental for a watertight copper sheath, the seam is checked in-line after welding, e.g. by eddy current measurements.

### 5.1.5 Lay-up

For the production of three-phase cables, the three cable cores must be laid up to form a coherent cable. Just taking three cores in parallel into a common armoring would render a very stiff design without flexibility. Horizontal lay-up machines have three (or more) pay-off drums containing the cable cores, and a take-up drum to receive the laid-up cable. The cable cores travel through a common die where they are put together and secured with a binder tape. The take-up drum is inserted in a drum-twister, which spins the drum axis around the cable axis while the take-up drum rotates and takes up the three-core cable. The lay-up is caused by the drum-twister movement. Also, the pay-off drums rotate in drum-twisters, providing the back-twist that is necessary to keep the three cable cores together. This is the same process as being used when making ropes.

This method is sufficient for short cables to be delivered on drums (probably less than 1000 m per length). For longer submarine cables it would be tedious to manufacture the cable in drum lengths and then joint the cables to achieve delivery lengths of 20–40 km. Such long lengths of three-phase submarine cable can be produced in vertical lay-up machines. The pay-off reels (drums or baskets) are mounted onto a turntable rotating around a vertical axis. The cable cores are fed vertically upward into a collecting die many meters above the turntable centre. The cable cores are pulled through the die, then over a sheave and then placed onto a take-up turntable. Figure 5.5 shows a vertical lay-up machine with rotating baskets as pay-off.

The baskets for the cable cores can accommodate some kilometres of cable core depending on the core diameter and basket size. When the baskets run empty, they are reloaded with the next cable core, which is jointed into the foregoing cable cores by means of flexible factory joints. The turntable carrying the baskets can also carry additional baskets or reels for optical cables and/or filler elements, which would be arranged in the interstices between the cable cores. Vertical cable lay-up machines are the most versatile lay-up machines.

Some of the difficulties caused by the three-phase lay-up topology can be overcome with SZ-lay-up. In this method the lay-up direction (right-handed or left-handed) is altered regularly. This method requires no drum twisters or rotating take-up reels.

**Fig. 5.5** Vertical lay-up machine with three rotating pay-off baskets and a fourth rotating pad for an optical cable. Additional small drums can accommodate filler ropes (Courtesy of ABB, Sweden)

The diameter of the envelope circle of the three-core cable is 2.16 times that of the individual cable cores.

### *5.1.6 Armoring*

The most prominent attribute of submarine power cables is the armoring. Basically, submarine cable armoring is the winding of metal wires around the single core or a multicore cable. The most common wire is galvanized steel wire, but copper, brass, bronze, and aluminium wires sometimes also are used. The beautiful engraving in Fig. 5.6 from 1898 illustrates the working principle of an armoring more clearly than most photographs [5]. Today's armoring machines are very similar except for electronic drives and computerized monitoring equipment. Armoring machines carry a number of wire bobbins arranged on a rotating cage. The cable is travelling through the centre axis of the armoring machine. As the cage rotates the wires are drawn off the bobbins into a collecting bellmouth die at the exit of the armoring machine. In the bellmouth the wires are laid around the cable in the centre. The rotating speed of the machine and the cable travel speed must be synchronized to maintain a correct

**Fig. 5.6** Armoring machine with planetary movement of the wire bobbins. Six bobbins in two groups are visible. The eccentric ring in front of the first rim keeps all bobbins in the same orientation during their rotation (Machine designed and produced by Johnson & Phillips, London)

lay length. Some submarine cable factories have two armoring machines in line to produce submarine cables with double armoring, often with opposite lay direction.

In "rigid" armoring machines the bobbin axles are set up rigidly in the cage frame which means that the axles turn around with the cage. Each wire receives a 360° twist for each rotation of the cage. This can be difficult with heavy wires and a short lay-length. In most large armoring machines the bobbins are guided by a planetary gear, which gives them the same global orientation on their travel around the cable (cf. Fig. 5.6). By this means, the armoring wires will not receive a twist as they are being laid around the cable. However, when flat armoring wires are to be used, the armoring machine must have no planetary movement. There are armoring machines where the planetary movement can be switched on/off at will. Each bobbin has a breaking device that keeps the wires straight, and provides a smooth laying onto the cable core.

Cable caterpillars or other linear cable engines situated downstream pull the cable through the armoring machine. Considerable forces are required to draw off the cable from the armoring machine since a multitude of armoring wires have to be drawn off the bobbins and wound around the cable. For large armored cable the necessary pulling force can be 5 tons or more.

The lay length of the armoring is determined by the ratio of cable speed through the machine and the rotation speed of the cage. The pull-of engine can be coupled mechanically to the rotation drive by means of gearboxes. Modern armoring machines with electronic d.c. drives can achieve any desired lay length with high accuracy.

Large armoring machines for submarine cables have often two rotating cages in a row. Very large submarine power cables can be covered with a complete layer of wires using both cages in tandem operation. The cages are rotated in the same direction with identical lay length, and the wires are combined to a single wire layer. For the production of a double layer uni-directional armoring the cages are rotated at different speeds so that the layers have different lay lengths. Much more common is a double-layer counter-helical armoring using the two cages in opposite rotation direction.

At regular intervals the armoring process must be stopped to swap the empty bobbins for refilled ones. The new wires are butt-welded to the trailing ends of the previous ones. This is a rather quick process, completed by de-burring and zink-coating of the weld. Swapping all bobbins of a machine cage at the same time keeps the downtime shorter. For cables intended for large depth it may be considered to spread the welds over a longer distance.

One or more bobbins can carry optical fibre (OF) cables to be integrated into the power cable. OF cables are more delicate than the rugged steel wires and require more attention, more sophisticated braking units and gentler internal guiding in the armoring machine.

Tape winder units are often situated at the entrance of the armoring machine to apply bedding tapes. Often also metallic tapes are applied under the armoring wires. They provide pressure reinforcement for paper-insulated cables, or a teredo protection where it is required.

The corrosion protection of the armoring is applied in the same process. Most often the corrosion protection is accomplished by flushing the armoring with fluid bitumen. The heated bitumen is showered over the armoring wires right before they are collected into the die, and again after the die. Using this arrangement the wires are covered with bitumen from all sides. The flow of bitumen must be cut off when the armoring machine is halted for wire change or other service, otherwise the cable under the bitumen shower would be overheated.

An outer serving of polypropylene threads (yarns) is finally wound over the armoring wires, one layer right-hand and one left-hand. The rotating stand for the thread spools is shown in Fig. 5.7. The bituminised armored cable can be seen entering into the centre orifice. The serving threads are immersed into the adhesive bitumen. The bitumen under the outermost serving layer, if any, should be applied sparsely, in order to avoid the bleeding of the cable during storage and laying.

Long submarine power cables often consist of a number of factory lengths, which are shorter than the shipping length. When a factory length of unarmored cable core is used up in the armoring process, the next factory length is pulled to a jointing room. The next length is then jointed to the previous one under controlled humidity conditions and high cleanliness requirements. Upon completion the joint is pulled through the armoring machine, and is covered by an uninterrupted wire armoring. The factory joint is only a few millimetres larger in diameter than the cable and is considered an integral part of the cable. Depending on the particular factory jointing technique this may require a 1–5 days break.

It is not possible to combine the cable armoring with the manufacturing of an extruded outer sheath as extrusion lines cannot be stopped easily.

**Fig. 5.7** A serving of polypropylene threads is applied on an armored cable covered with hot bitumen (Courtesy of ABB, Sweden)

### *5.1.7 Storage of Submarine Cables*

It can be worthwhile to make some comments on the storage of submarine cables because it may involve the storage of a single product of 7000 tons in weight, without having the means of lifting the entire thing.

Cable drums can carry some hundred meters of armored submarine cables, supersize drums possibly 1–2 km. This storage form is suitable for in-field cables for OWPs, for shorter beach cables and limited water crossings. Cable drums with armored cable can have a weight of 30–50 tons and require suitable lifting and transport gear, and driveways with sufficient stability.

Longer lengths would be stored on turntables or fixed coiling pads. A coiling pad is a flat round or oval surface on the factory premises or in a storing tank on-board of the cable ship. The cable is guided down onto the pad from an elevated position, the guiding mechanism often equipped with a linear cable engine (cf. Fig. 5.8).

Coiling of armored cable can only be done if the armoring is unidirectional. Each turn of the cable around the coiling pad will cause a 360° twist in the cable. Therefore a minimum coiling diameter is required to enable the cable to absorb the twisting generated in each turn. The minimum diameter of the coiling ring should be no less than 60 times the cable diameter [6]. The exact minimum coiling diameter is depending on the armoring lay-length, the bedding under the wires and other design parameters.

**Fig. 5.8** Coiling of submarine power cable

A cable with uni-directional (single layer or double layer with the same lay direction) armoring can absorb torsions only in one direction. Coiling is only possible if the cable is twisted in the direction where the armoring is opening up. Coiling in the wrong direction will inevitably lead to kinks and loops in the cable and the entire procedure will fail.

For an S-lay cable the coiling direction must be clockwise. Under clockwise coiling the armoring can open up a little under the influence of the twist in the cable. Choosing the wrong direction of coiling (counterclockwise for a cable with S-laid wires) would provoke a closing-in of the armoring wires. Under these forces the cable would tend to form kinks or loops. However, as an exception to this rule, 3C-cables with single wire armoring and the cable core lay-up in opposite direction can be laid up counterclockwise (cf. Table 5.1).

The coiling pad should be flat. A high fence around the coiling pad provides protection to the coil staff that walks around the coil to direct the cable in position. The cable is fed to the coil from a point some 6–15 m above the coil. It is of advantage to have a caterpillar or wheel engine on the top. A fenced coiling pad with five coilers is shown in Fig. 5.8. The quadrant (roller guide arc) is suspended from a crane and contains a wheel-pair cable engine.

Direct solar radiation on the stored cables can develop high temperatures that easily can unpleasantly bleed out the bitumen.

Cables on coiling pads, outdoor turntables, trailers or cable laying vessels are frequently showered by rain, seawater spray, overflows and similar. In these situations

**Table 5.1** Coiling direction depending on armoring lay orientation

| | Armoring lay orientation | Coiling |
|---|---|---|
| Single core cables with unidirectional armouring | S-lay | Clockwise |
| Single core cables with counter-helical armouring | S or Z | No coiling |
| 3C cables with core unidirectional armoring, core lay-up in S-orientation | S-lay | Clockwise |
| 3C cables with core unidirectional armoring, core lay-up in Z-orientation | S-lay | Clockwise or counterclockwise |

cable ends must always be capped with suitable end caps recommended by the cable supplier. Shrink-on end caps are in most cases not advisable.

## 5.2 Testing

Submarine power cables are subjected to comprehensive tests during development, qualification, manufacturing and installation. The various tests serve different purposes with the single overall goal – to ascertain a trouble-free operation under specified conditions.

The large variety of cable related tests can be categorised according to their purpose and stage during qualification, manufacturing and installation of the cables.

### *5.2.1 Development Tests*

The development of new cable types, or the extension of existing cable types to new sizes or ratings, may require comprehensive testing of materials, components, and production processes. Many of the new materials and concepts have been developed for underground cables before they were being employed for submarine cables. The reason is simply that possible failures in submarine cables cause much higher repair costs and longer outage time compared to identical failures in underground cables.

The polymeric components of cables are subject to continuous improvement. New polymer formulations proposed by material suppliers need to be evaluated under the specific conditions of submarine power cables. Tests of dielectric strength, loss angle, leakage current, and dielectric response test, are used to characterize and screen insulation materials. These tests can often be performed on samples or model cables avoiding the expensive manufacturing of full-size cables. The ageing performance is being tested under different temperature and ambient conditions. The processing of cross-linked materials can be evaluated with methods determining the cross-linking degree.

Polymeric sheath materials in submarine cables are sometimes tested for their water vapour permeability, stability in salt water, abrasion resistance, carbon-black content, etc. However, in most cases manufacturers use their well-proven sheath materials without the need for change.

Even if most properties of metallic constituent materials are listed in textbooks, some properties need to be evaluated. The fatigue properties of lead sheathing alloys must be assessed by testing very thoroughly, taking the manufacturing method into account. Other metallic sheathing materials are also subjected to fatigue tests to secure trouble-free performance during the cable lifetime.

Another group of development tests addresses the corrosion performance of all metallic materials in submarine cables. Metals are tested separately, but also in relevant combinations to investigate galvanic behaviour.

Mechanical tests on complete submarine cables are designed to demonstrate their ability to withstand all stresses and incidents, which can be encountered during installation and operation of the cable. These tests include tensioning, bending, pinching, impacts, etc. Also the behaviour under the influence of external hydrostatic pressure is important for submarine cables.

Knowing the strong and sensitive points of the cable design, and knowing which tests will be required for a successful qualification, every manufacturer has his own set of tests to be pursued during development programs. For a speedy screening, sometimes short-cut tests are designed that can deliver reasonably reliable results in short time. This is of particular interest for tests for the ageing behaviour of cable materials. The cable maker would subject the newly developed cable design to these pre-tests, before he brings it to the standardised qualifying long-term test.

### *5.2.2 Type Tests*

Once a cable type has been developed or adopted to new applications, it will be subjected to a type test. As many large submarine cable projects require a tailored unique design, many purchase contracts also require the performance of type tests. However, as type tests add considerably to costs and project execution time it should always be considered if the proof that a certain cable design is suitable for the intended project, could be deducted from previous type tests on similar cable types.

The purpose of the type test is to "qualify the design and the manufacturing of the cable system against the conditions of the intended application" [7]. Type-testing of electric equipment is generally regulated by test standards issued by national authorities or professional organisations such as IEEE, AEIC, ANSI, or Cigré. Most power cable testing standards cover underground land cables only, and some standards explicitly exclude submarine power cables from their scope.[3] Therefore test standards for submarine power cables are scarce (Table 5.2).

In practical life, type-tests on submarine power cables are often performed according to test standards applicable to underground cables with the same insulation design and conductor size. These type tests include material tests and electric tests, and in some cases mechanical tests such as bending tests over a cable drum.

[3]Cables for submarine applications are explicitly excluded from the scope of IEC 60502-2. However, it can be considered to use IEC 60502-2 for informational tests in medium voltage submarine cables.

**Table 5.2** Type test standards usable for submarine power cables

| | Published in | Title or content |
|---|---|---|
| Cigré | Electra No. 171 April 1997 | Recommendations for Mechanical tests on sub-marine cables<br>Referred to as Electra 171 in the following |
| Cigré | Electra No. 189 April 2000, pp. 29ff [8] | Recommendations for testing of long a.c. submarine cables with extruded insulation for system voltage above 30 (36) to 150 (170) kV<br>Referred to as Electra 189a in the following |
| Cigré | Electra No. 189 April 2000, pp. 39ff [7] | Recommendations for tests of power transmission dc cables for rated voltages up to 800 kV (all insulation types excl. extruded)<br>Referred to as Electra 189b in the following |
| Cigré Technical Brochure TB 219 | Cigré Technical Brochure 219, Working group 21.01, February 2003 [9] | Recommendations for testing DC extruded cable systems for power transmission at a rated voltage up to 250 kV<br>(scope includes submarine cables) |
| IEC 60840 | | Power cables with extruded insulation and their accessories for rated voltages above 30 kV ($U_m = 36$ kV) up to 150 kV ($U_m = 170$ kV) – Test methods and requirements |
| IEC 62067 | | Power cables with extruded insulation and their accessories for rated voltages above 150 kV ($U_m = 170$ kV) up to 500 kV ($U_m = 550$ kV) – Test methods and requirements |

The Cigré recommendation published in Electra 171 is the only known test standard describing relevant mechanical tests on submarine power cables. Sometimes the type test specifications include mechanical tests according to the Electra 171 followed by applicable type test elements from IEC standards (for a.c.) or Cigré recommendation (for d.c.). It shall be noted that type tests are normally not mandatory by legal means. The test conditions are sometimes part of the contract negotiations.

Table 5.3 lists the applicable type test standards for the five generic cable types as described in Chap. 2. These standards are adhered to in most European submarine cable projects. Different type test standards may be applicable in other markets.

#### 5.2.2.1 Mechanical Tests

The Cigré test recommendation published in Electra No. 171, April 1997, suggests a few test procedures especially designed for submarine power cables. It is at the discretion of the purchaser to determine which tests to be done.

**Table 5.3** Electric type tests for five generic cable types

| Cable type No | 1 | 2 | 3 | 4 | 5 |
|---|---|---|---|---|---|
| Rated voltage $U_0$ | 33 kV a.c. | 150 kV a.c. | 420 kV a.c. | 150 kV d.c. | 450 kV d.c. |
| Insulation | XLPE | XLPE | Paper/oil | Polymer | Mass-impregnated |
| Mechanical tests | | | Electra 171[4] | | |
| Electrical type test | Electra 189a with reference to IEC 60840 | IEC 60840 | IEC 62067 | Cigré Technical Brochure TB 219 | Electra 189b |
| Test sequence | TB<br>PD<br>Tan δ<br>HC<br>PD<br>LI<br>AC<br>(PD)<br>(same as in IEC 60502) | TB<br>PD<br>Tan δ<br>HC<br>PD<br>LI<br>AC<br>(PD) | TB<br>PD<br>Tan δ<br>HC<br>PD<br>SI (for $U_m \geq 300$ kV)<br>LI<br>AC<br>(PD) | TB<br>LC<br>SI<br>LI & DC<br>DC | TB<br>LC<br>PR<br>LI<br>SI |

Abbreviations: TB, Tensile bending test; WP, Water penetration test; PD, Partial discharge test; HC, Heating cycle voltage test; LI, Lightning impulse test; SI, Switching impulse test; LC, Load cycle test; PR, Polarity reversal test; DC, High-voltage test with d.c.; AC, High-voltage test with a.c.

The first suggested test is a coiling test, to be performed only on those cables that are intended to be coiled during manufacturing, loading, or installation. The test confirms the suitability of the cable for coiling according to the parameters encountered in cable handling. The coiling diameter is not specified in Electra 171 but the test shall confirm the value given by the manufacturer. The test cable must be long enough to achieve a sufficient number of turns to exclude end effects.

The tensile bending test according to Electra 171 demonstrates the cable's ability to withstand tensional forces in combination with bending over the laying wheel during installation. The test is to be performed on a piece of cable before the electric tests are performed on the same piece of cable. A flexible joint can be qualified when included in the test cable. The test cable is laid halfway around a large sheave, which represents the laying wheel of the cable-laying vessel. A pulling force is applied to the cable ends while they are moved three times back and forth around the wheel. The wheel diameter is often 6 or 10 m representing typical laying wheel diameters of existing cable ships (cf. Fig. 5.9).

[4]Electra No. 171 is "primarily meant for cables having a rated voltage $U_0$ higher than 36 kV a.c. or 100 kV d.c.". However, Electra 189a covering 33 kV cables refers to Electra 171.

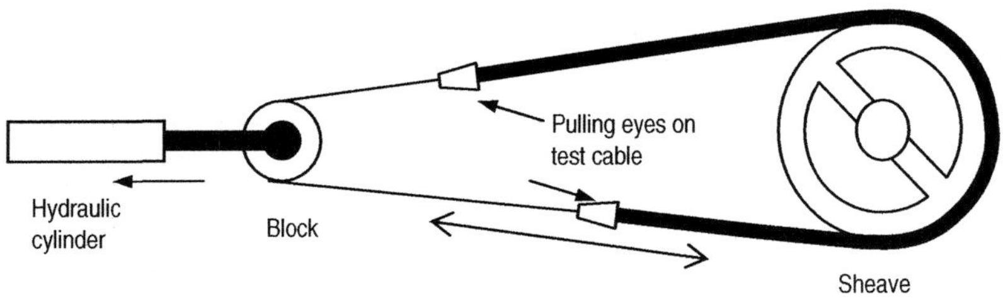


**Fig. 5.9** Test arrangement for tensional bending test according Electra No. 171. *Top view*

The test pulling force is given in Electra 171 depending on the cable weight in water and the water depth (cf. Chap. 3). For a depth over 500 m, also the dynamic forces caused by wave movements are included. Experience has shown that actual pulling forces on the laying wheel can considerably exceed the forces given in Electra 171, depending on the weather situation during laying.

For many submarine transmission projects the laying depth is below 100 m, which leads to rather moderate test forces according to the Cigré recommendation. Sometimes the required pulling force is so small that the heavy cable is not even straight but has considerable sag during the test. Tests with the Cigré-recommended forces for laying depth under 100 m often have no practical relevance. Purchaser and manufacturer should consider an agreement to omit the test to save costs.

The tensional bending test confirms the design values of tensional strength and side-wall pressure (SWP), i.e. the uniform lateral force onto the cable side per unit length.[5] Other mechanical stresses might be experienced during installation, such as localized side impact by caterpillar pads or wheeled cable engines, or stresses caused by trenching machinery. Also the transport of cable over curved roller tracks or cable gantries is not covered by the Cigré test recommendations, unless the rollers in the gantry are very close together and the roller diameter is sufficiently large. No type tests have been devised yet for this type of impacts. As installation conditions and gear vary largely between projects, it would not be possible to construct type tests covering all type of impacts. Instead the manufacturer, purchaser and installer should agree on the necessity and character of additional confirming tests.

The Electra No. 171 also recommends an internal pressure test for oil-filled and gas-filled submarine cables.

A very particular test recommended in Electra No. 171 is a sea trial test for cases where laying conditions differ from previously performed operations. For a sea trial test the anticipated laying spread (vessel, on-board equipment and crew if possible) should be established to lay and recover a substantial length of the submarine cable on a seafloor representative for the target area. The tests are very expensive but can reveal possible flaws in cable design and installation equipment. Poor equipment or a cable design that is not appropriate for the installation method can cause immense costs when discovered during the cable installation. In the worst case, the instal-

[5]The established expression "side-wall pressure" actually does not denote a pressure ($N/m^2$) in a classical sense but a force per unit length. The unit of SWP is kN/m.

lation must be interrupted and the equipment upgraded or refurbished. It is by far less expensive to do this in a well-planned manner after a sea trial test prior to the installation campaign. Acknowledging this, many major submarine power projects have performed sea-trial tests and used the results for the benefit of the project.

#### 5.2.2.2 Load Cycle Test

Load cycle tests are part of the type tests for all submarine power cables. The a.c. cables are usually subjected to electric tests according to the same specifications as underground cables. The load-cycle test procedures for d.c. cables are less well-known and shall be described in the following.

The recommended load cycle test for mass-impregnated HVDC cables is described in Electra 189b. A 24-hours load cycle is combined from an eight-hour period of full cable load, followed by a 16-hours period of natural cooling. The Electra 189b load cycle test comprises ten load cycles with a constant negative voltage of $-1.8 \times U_0$. After this, another ten cycles will be performed with positive voltage of $1.8 \times U_0$. The ambient temperature during the test is subject to agreement between manufacturer and purchaser. Depending on the conditions in the seafloor, the tests may be specified with cold and warm ambient to reflect different thermal conditions and seasons. For a cable installed in the seafloor in moderate waters, the lowest expected ambient is around 6°C when the cable has not been used for a while, and possibly over 30°C after a long uninterrupted period of power transmission. In tropical waters, the ambient temperature would be specified differently. The ambient temperature of the test cable can easily be adjusted if the cable is installed in hoses or pipes with temperature-controlled water. The heating current amplitude should reflect the rated load of the cable and is also subject to agreement.

The limiting factors for the operation of mass-impregnated HVDC cables are the conductor temperature and the temperature drop across the insulation. During type test, only three of the four following parameters can be set independently:

- Conductor temperature
- Temperature drop over the insulation
- Conductor current
- Ambient temperature.

The 8/16 h rhythm of the cycles is widely accepted as it fits well into the working day. As the thermal time constant of this type of cables is 1 to 2 h, a shorter cooling/heating cycle would also achieve appropriate heating of the cable. For development tests, a 4/8 h sequence is an acceptable short-cut.

The test voltage deserves some extra comments. Although the load cycle test for mass-impregnated HVDC cables is described in Electra 189b, its conditions are often altered during contract negotiations. Electra 189b specifies a test voltage of $1.8 \times U_0$, where $U_0$ is the operating voltage of the HVDC cables. In recent years manufacturers and cable purchasers have often agreed on lower test voltage factors, in particular for the cooling part of the load cycle test. The reason is simply that the dielectric strength during the load phase of the type test (with full current), and in most operational situations is much higher than in the type test cooling phase. It would be uneconomic to oversize an HVDC cable just in order to comply with a single test parameter. For mass-impregnated HVDC cables of the past 15 years a test voltage of $1.55 \times U_0$, $1.6 \times U_0$, or $1.7 \times U_0$ have been specified by purchasers for the cooling phase.

The test cable is terminated with oil-filled pressurized cable terminations. Oil inflow from these terminations into the cable must be prevented during the type test, as this would assist the cable to pass the type test, a situation not available in real life at some distance from the termination. Means to prevent this oil flow must be provided in order to assure relevant type-test conditions. This can be achieved by extra long cable connections between the terminations and the test cable portion, by freezing the cable between the terminations and the test cable portion, or by sealing arrangements inside the terminations.

The next step, after negative and positive voltage load cycles, is the polarity reversal test. The load cycle (8/16 h) is maintained but the voltage is periodically reversed between $+1.4 \times U_0$ and $-1.4 \times U_0$ every 4 h. This is repeated during ten complete load cycles. The recommended reversal speed specified in the Electra 189b is under 2 min, but real tests have been performed with longer reversal times, due to laboratory equipment reasons, or due to transmission system aspects not requiring such short reversal times [10]. After the polarity reversal tests the cable is subjected to impulse testing.

For extruded HVDC cables, which have conquered a reasonable portion of the submarine HVDC cable market, a load cycle test is specified in Cigré Technical Brochure 219 [9]. For cables to be qualified for voltage source converters, i.e. without polarity changes, TB 219 recommends the following load cycle test sequence (Table 5.4):

**Table 5.4** Load cycle test sequence for extruded HVDC submarine cables according to Cigré TB 219

| Load cycle characteristics | Number of cycles |
|---|---|
| 8/16 h load cycles at $-1.85 \cdot U_0$ | 12 |
| 8/16 h load cycles at $+1.85 \cdot U_0$ | 12 |
| 24/24 h load cycles at $+1.85 \cdot U_0$ | 3 |

For extruded HVDC cables to be qualified for line-commutated converters (so called "Classic"), a different suite of load cycles is defined, including polarity reversal tests. Owing to the characteristics of extruded d.c. insulation, a reduction of test voltage in the cooling phase as in the case of mass-impregnated HVDC cables has not been advocated yet.

TB 219 requires the performance of a row of non-electric tests before the load cycle tests. These tests comprise bending tests (depending on submarine or underground application) and water integrity tests, with reference to other IEC or Cigré test standards. After the load cycle tests the cable is subjected to impulse testing.

### 5.2.2.3 Impulse Tests

Impulse tests are a part of most type test standards. It became clear very early in power engineering that a.c. and d.c. tests alone do not cover all events in the life of a component in the power system. The dielectric strength of any insulation is very much depending on the shape and duration of a voltage wave. Impulse tests are necessary to demonstrate the behaviour of components under temporary overvoltages. The tests are performed according to IEC test standards, which apply in principle to all high-voltage power transmission equipment. As for other high-voltage cables, the test standards specify lightning impulse tests and often also switching impulse tests for submarine cables. Despite their names, the tests do not necessarily resemble real life switching or lightning impulses. The distinction of switching and lightning impulses may be made on the basis of wave shape rather than their origin [11]. The impulse shapes are specified in IEC 60060-1 in terms of front steepness and tail time to half value. The standard lightning impulse (LI) has 1.2 μs front time ($t_1$) and 50 μs time to half value ($t_2$) after the crest. For the standard switching impulse (SI) these values are 250 and 2500 μs, resp. IEEE defines 1.4/40 μs as LI wave shape parameters. IEC and IEEE use slightly different formulae as how to determine $t_1$ and $t_2$ from the wave shape. Figure 5.10 shows the ideal impulse shape. However, laboratory records usually do not look like this and the evaluation of the test parameters of a specific test (front time, time to half value) is all but easy. Computer-assisted algorithms help to identify parameters from corrupted records with distortions and overlaid ghost oscillations [12, 13].

Impulse test parameters were originally conceived to resemble phenomena in air-insulated power grids and substations. Today, the situation for many submarine cable terminations is different. HVDC cables are connected to converter stations where certain types of overvoltages cannot occur at all, or only with lower amplitude compared to air-insulated a.c. substations. Other modern submarine power cable applications, e.g. in HVDC converters or offshore, have their terminations indoors.

Type tests for a.c. submarine cables comprise impulse tests according to standards for land cables with corresponding rating. For d.c. cables the Cigré recommendations specify impulse tests with superimposed d.c. voltage.

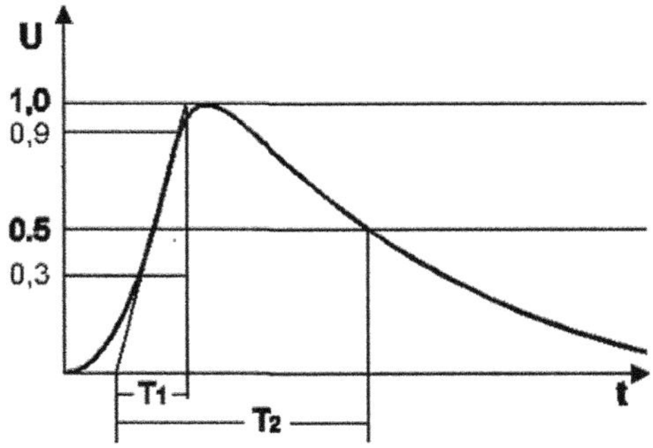


**Fig. 5.10** Standard impulse voltage shape

For paper-insulated HVDC cables ten positive lightning impulse shots superimposed over negative d.c. voltage are performed, then ten negative lightning impulse voltage shots superimposed over positive d.c. voltage. The same sequence is done with switching impulses. This concludes the electric type test for paper-insulated HVDC cables.

According to the Cigré TB 219, extruded submarine HVDC cable systems have to undergo a complicated pattern of switching impulse tests with opposite or parallel polarity. Lightning impulse testing may be omitted under certain circumstances.

## *5.2.3 Routine Tests*

Routine tests are performed on all manufacturing lengths and/or delivery lengths. In the various test standards applicable for submarine power cables, different sequences of routine tests are prescribed. Almost all standards require testing of conductor resistance, insulation capacitance, and loss angle tan $\delta$. Resistance and capacitance can be measured either on the complete cable (e.g. on a drum length), or on end pieces cut from a long length of cable. The resistance measurement requires the knowledge of the conductor temperature. For short test pieces, a four-point resistance measurement should be used. The distance between current source contacts to the conductor and the voltage probes should be sufficiently long to allow for a uniform current distribution in the conductor. Should the resistance test for a manufacturing length in a long cable exceed the required value, it should be checked if there are expected hotspots in the cable route sector for which the manufacturing length is destined. If not, the total resistance of the cable route can be adjusted by adding a little to the conductor area of the remaining manufacturing lengths.

The capacitance test is a check that the insulation is not too thin. According to IEC standards the specified value must not be exceeded by more than 8%.

The loss angle tan $\delta$ is conveniently measured on a piece of cable cut from a manufacturing length. The IEC standards require the measurements under a.c. voltages corresponding to an electric stress in the insulation of 2 and 8 kV/mm, resp. The measurement can be performed e.g. with a high-voltage Schering bridge. An elaborate description of tan $\delta$ measurement methods and their theory can be found in [1]. Limit values are specified both for tan $\delta$ at both voltages, and for the difference between both results. The American AEIC standards specify the ionisation factor, IF, being the difference between tan $\delta$ at 100 V/mil (=3.9 kV/mm) and 20 V/mil (=0.8 kV/mm). The IF value must not be higher than 0.1% for LPOF cables.

High tan $\delta$ values can hint on contaminated insulation, which leads to high ohmic leakage currents through the insulation. The tan $\delta$ value can also be increased by partial discharges in the insulation, resulting in an integral current.

### 5.2.3.1 High-Voltage Routine Tests

Manufacturing lengths are often given a high-voltage routine test either by standard requirement, by agreement, or as an internal test of the manufacturer (Table 5.5). The obvious reason is to exclude faulty cable core from further costly manufacturing

**Table 5.5** a.c. test voltages for routine and after-installation tests on a.c. cables

| | Rated voltage kV | $U_0$ kV | Routine test voltage (phase-to-ground), kV | After installation test voltage (phase-to-ground), kV |
|---|---|---|---|---|
| IEC 60605 | 30 | 18 | 63 | 30 (5 min) *or* 18 (24 h) |
| IEC 60840 | 45–47 | 25 | 65 | 52 |
| | 60–69 | 36 | 90 | 72 |
| | 110–115 | 64 | 160 | 128 |
| | 132–138 | 76 | 190 | 132 |
| | 150–161 | 87 | 218 | 150 |
| IEC 62067 | 220–230 | 127 | 318 | 180 |
| | 275–287 | 160 | 400 | 210 |
| | 330–345 | 190 | 420 | 250 |
| | 380–400 | 220 | 440 | 260 |
| | 500 | 290 | 580 | 320 |

steps. The voltage factors given in standards are different for different cable types, and can also be altered in agreement with the client.

Modern submarine power cable cores can be manufactured in appreciably long length of 20 km or more. Factory testing of these long cable cores with a.c. voltage requires considerable testing power to provide the charging current. In many cases, this is beyond the possibilities of a.c. test voltage generators. More testing power can be achieved with resonance circuits where an inductance is tuned to form a resonance circuit with the test cable capacitance. The testing frequency of resonance systems is between 30 and 300 Hz.

Testing with d.c. voltages (e.g. HVDC cables) requires the charging of the capacitance of the cable up to the specified voltage. The charging current of d.c. test voltage generators is limited and it may take a considerable time to reach the desired voltage. The charged cable capacitance stores a large amount of energy. Strict fencing of the test range should be provided for the case that the energy is released in a breakdown. Should a test termination fail the amount of stored energy can easily explode a porcelain insulator.

### *5.2.4 Factory Acceptance Tests (FAT)*

FAT is the last test program to be performed at delivery. The performance of FAT is often connected to the issuing of client certificates and project milestones. What has been said about testing methods in the routine tests, also applies to the FAT. As delivery lengths can be much longer than the included manufacturing lengths the charging time for d.c. tests are even longer. Also the discharge of the tested cable after the FAT takes considerable time. In case of a.c. FAT on long submarine cables a sizeable set-up of resonance circuit has to be employed (cf. Sect. 5.2.5). Often only specialized submarine power cable makers can provide this type of FAT.

The American standard ICEA No. S-57-401/NEMA Standards Publication No. WC2 recommends a high-voltage a.c. test for armored cables employing 80% of the test voltage for an equivalent un-armored cable.

The FAT for long submarine cables may even include a TDR (Time Domain Reflectometry). The test comprises sending a short voltage impulse into one cable end and recording the echoed impulse. The travelling impulse is partly reflected by changing of $Z$, the impedance per unit length of the cable. The most prominent change of impedance is of course the far end of the cable where the impulse is being reflected. Also factory joints sometimes constitute a local change in $Z$ causing a faint partial reflection of the travelling impulse. The idea behind the TDR test is to create a reference curve, which can help localising possible future cable faults. There are no pass/fail criteria for the TDR test.

### 5.2.5 After-Installation Test

A damage-free installation of the submarine cable is confirmed by a successful after-installation test of the entire cable link including joints and terminations (Table 5.6). As in FAT, the tests are performed as high-voltage test with either a.c. or d.c. It is not always easy to bring high-voltage sources to the installation site.

d.c. test voltage generators can be transported quite easy in a truckload. Since the test voltage generators are mostly designed for indoor laboratory use, they must be used only indoors or under dry weather conditions. Fog and dew can degrade the internal insulation of the test equipment. Sometimes temporary shelters can be erected to protect the equipment. Especially with long cable links the testing time can be long due to the cable charging. In one of the recent very long HVDC projects the specified testing time at $1.4 \times U_0$ was only 30 min, but the complete testing took 11 hours due to charging/discharging time. About five hours of this time the voltage was above $U_0$.

**Table 5.6** Test voltages for routine tests and after-installation tests

| Cable type No | 1 | 2 | 3 | 4 | 5 |
|---|---|---|---|---|---|
| Rated voltage $U_0$ | 33 kV a.c. | 150 kV a.c. | 420 kV a.c. | 150 kV d.c. | 450 kV d.c. |
| Insulation | XLPE | XLPE | Paper/oil | Polymer | Mass-impregnated |
| Applicable standard | Electra 189a with reference to IEC 60840 | | IEC 62067 | Cigré TB 219 | Electra 189b |
| Routine test voltage | – | 218 kV | 440 kV | $1.85 \times U_0 =$ –278 kV d.c. | $1.8 \times U_0 =$ –810 kV d.c. 15 min |
| After-installation test voltage | – | 150 kV | 260 kV | $1.45 \times U_0 =$ –218 kV d.c. 15 min | $1.4 \times U_0 =$ –630 kV d.c. 15 min |

Testing long cable links with high-voltage a.c. requires substantial charging currents. The required power for the test circuit is proportional to the cable capacitance and the square of the test voltage. For many cable links, the a.c. testing would require a test power exceeding the capacity of most high-voltage test transformers. Another strategy to test a.c. cables is to connect them to the power grid for 24 h. Obviously, this method is independent on heavy test transformers. The method is also called soak test and is applied frequently to h.v. a.c. cables. The test method has also been recognized in IEC 60840. However, the soak test does not provide testing at elevated voltages. Another strategy is using d.c. voltage also for testing of a.c. cables. This enables the use of simpler test equipment. However, it is not sure that hidden flaws in the cable system can always be detected by d.c. voltage. Even worse, many specialists agree that the d.c. voltage can do more harm than good, and deteriorate the insulation. Today, d.c. testing on a.c. submarine cables is used only to a lesser extend.

Long and very long cable links can be tested at elevated voltages with series resonance circuits. In these test circuits an inductance is connected to the cable capacitance to form a resonance circuit. An exciting transformer supplies the voltage to the resonance circuit in which an a.c. voltage can be maintained by resonance action between the cable capacitance $C_{\text{cable}}$ and the external inductance $L_{\text{external}}$. The resonance criterion is:

$$C_{\text{cable}} \cdot L_{\text{external}} = \frac{1}{(2\pi f)^2}$$

where $f$ is the power frequency. In some commercial series resonance circuits the resonance criterion is met by a tunable external inductance at constant frequency, while other systems employ a tunable frequency at constant external inductance. The tunable frequency systems are said to have a better power/weight ratio, a fact that can be very important given the weight of the necessary inductance [14, 15].

The power frequency is in the range of 30–300 Hz. Some commercial resonance testing systems are built modular, i.e. more inductances can be added in order to meet the larger cable capacitance of very long cable links. A wider frequency range is being discussed for future test standards.

The resonance circuit made up of the cable and the external inductance is lossy. In every oscillation of energy between the capacitance and the inductance, a certain amount of energy is dissipated in the ohmic components of the circuit, mainly the cable conductor and screen. This ohmic loss must be replenished by the exciter transformer, and can limit the maximum length of the test cable circuit. A cable length of well above 100 km can be tested, depending of cable capacity, test voltage, system configuration, and desired test duration. Suppliers of test systems can provide exact data in their system specifications.

Depending on the needed a.c. test voltage and the capacitance of the cable, one or more truckloads of equipment must be shipped to the testing site. The method is feasible for submarine cable links where at least one end is accessible easily from a landing site. However, very large submarine cable would require unreasonable amounts of combined inductances to perform the task.

The American standard ICEA No. S-57-401/NEMA Standards Publication No. WC2 recommends a high-voltage after-installation test for armored cables employing 80% of the test voltage used for the test on the armored cable in the factory according to Sect. 5.2.4, i.e. 64% of the factory test voltage for an equivalent un-armored cable.

### *5.2.6 Non-electrical Tests*

The various test standards stipulate, beside the electric tests, also a number of non-electric tests. The tests are performed as sample tests on a defined share of the entire production, as routine tests on the entire production for a certain order, or as a part of the type test. The purpose of these tests is generally to confirm that the physical properties of the produced cable comply with the specifications.

A large spectrum of different non-electric tests is specified in the various test standards, to be performed at very different frequencies and occasions. It is almost impossible to compile a comprehensive summary on this item.

Many non-electrical tests are dimensional checks such as layer thickness or eccentricity. Other test specifications deal with the electric resistivity of semi-conducting materials, or simply the counting of the wires in the conductor. Material properties are checked before and after ageing. Hot-set tests on cross-linked materials check the quality of cross-linking.

Some of these tests are specified in the international test standards, others are specified by utilities, or just common practice in certain countries. The non-electric tests are often in the shadow of the more prominent electric tests. Because of this and the fact that the large number of different test specifications is somewhat confusing, it is recommended to define the bouquet of non-electric tests very clearly in the project contract.

## References

1. Bartnikas R (Ed.) (2000). Power and Communication Cables, IEEE Press Series on Power Engineering, ISBN 0-7803-1196-5.
2. McAllister D (1982). Electric Cables Handbook. Granada Technical Books Ltd, 1982, ISBN 0-246-11467-3.
3. Goodwin F E et al. (1994). Recent Technological Developments for Lead Sheathed Cables. Transmission and Distribution Conference, 1994. Proceedings of the 1994 IEEE Power Engineering Society, 10–15 April 1994, pp. 568ff, Chicago, USA.
4. Worzyk T (1996). Keine Gefahr bei Lecks in HGÜ-Seekabeln. Elektrizitätswirtschaft, Vol. 26.
5. Wilkinson H D (1898). Submarine Cable Laying and Repairing, The Electrician Printing and Publishing Company, London.
6. Gilbertson O (2000). Electric Cables for Power and Signal Transmission, John Wiley & Sons, New York.
7. Recommendations for Tests of Power Transmission dc Cables for Rated Voltages up to 800 kV. Electra No. 189, April 2000, pp. 39ff.

8. Recommendations for Testing of Long ac Submarine Cables with Extruded Insulation for System Voltage above 30 (36) to 150 (170) kV. Electra No. 189, April 2000, pp. 29ff.
9. Cigré publication of Working group 21.01 (2003). Recommendations for Testing DC Extruded Cable Systems for Power Transmission at a Rated Voltage up to 250 kV. Cigré Technical Brochure 219, February 2003, Paris, France.
10. Bex S et al. (2006). Basslink HVDC Design Provisions Supporting Ac System Performance. Cigré session 2006, Paper B4-301.
11. Agrawal K C (2001). Industrial Power Engineering and Applications Handbook, Newnes, ISBN 0750673516, 9780750673518.
12. Hällström J and Bergman A et al. (2007). International Comparison of Software for Calculation of Lightning Impulse Parameters Based on a New Processing Algorithm. Proceedings of the XVth International Symposium on High Voltage Engineering, Paper T10-539.
13. Berlijn S et al. (1999). Today's Problems with the Evaluation Methods of Full Lightning Impulse Parameters as Described in IEC 60060-1, 11th International Symposium on High Voltage Engineering, 1999. Conferences Publ. No. 467, Vol. 1, pp. 49–52.
14. Schufft W et al. (2000). Resonant Test System with Variable Frequency for On-site Testing and Diagnostics of Cables, HV Testing, Monitoring and Diagnostics Workshop, Alexandria, VA, USA, 13–14 September 2000.
15. Schikarski P et al. (1999). Two Years of Experience with a Mobile Resonant Test System for Testing of Installed Medium- and High Voltage Power Cables, International Symposium on High-Voltage Engineering, London, UK, August 1999.

# Chapter 6
# Marine Survey

## Contents



A sufficiently detailed marine survey is one of the most important activities during the planning and preparation of a submarine power cable project. The survey can deliver critical input data for the choice of the best cable route, the cable design, and the design of the cable installation procedure. These parameters do not only have an important impact on cost and schedule. A comprehensive survey can also make a substantial difference for the future cable link availability and repair costs. The profitability of the cable link can critically depend on the survey.

Based on the survey data of the submarine area between the landing points, the best possible cable route can be suggested in order to avoid submarine obstacles and hazards, such as:

- Existing cables and pipelines
- Fishing and trawling
- Shipping lanes, harbour entries, anchorages
- Dumping and dredging
- Seabed contamination
- Oil and gas fields
- Military activities
- Ammunition dumping grounds
- Ship wrecks and other submarine junk, abandoned cables
- Areas of sand or gravel extraction.

T. Worzyk, *Submarine Power Cables*, Power Systems,
DOI 10.1007/978-3-642-01270-9_6, © Springer-Verlag Berlin Heidelberg 2009

Troublesome areas should be avoided, since every major obstacle to a smooth laying operation can generate large costs. Avoiding these areas can be an important piece in the protection concept for the cable project.

The best cable route is also depending on the knowledge of the seafloor conditions and properties. In particular, the following data are necessary to know:

- Bathymetry, i.e. the water depth along the entire cable route
- Seafloor structure, existence of boulder fields, outcrops, crevices, canyons, steep slopes, rocky irregular sea floor
- Areas with risk for free spans of the cable
- Seafloor morphology, i.e. the structure of the soil beneath the seafloor
- Geophysical data such as grain size, hardness, thermal conductivity
- Sediment movements.

A thorough knowledge of the seabed conditions helps the project engineers to identify and locate potential hazards to operations. The data has impact on the cable design and the choice of the installation spread. If the seafloor conditions are too severe along the suggested cable route, this might be reason enough to change it.

Hydrological data also have a strong influence on cable design and installation procedures:

- Water temperature (surface and seafloor) over the year
- Wave patterns
- Tides and currents
- Wind patterns over the year.

Wind, waves, and tides have the ability to seriously hamper or interrupt an installation campaign so it is good to know in advance where the limitations are.

Other items may not endanger the cable but may be disturbed by the cable laying action:

- Bird and fish nurseries
- Sensible natural resources
- Tourist resorts.

The laying operation in or close to these locations is usually restricted or regulated by authorities. In some countries, the authorities might be weak or do not care. It is a matter of corporate ethics and environmental policies to decide what to do and not to do.

## 6.1 Scope of the Marine Survey

There are no strict general rules for the needed scope of a marine survey. It should be clear to the submarine cable project planner that a more comprehensive survey can help the installing contractor to perform a more precise tender and a better installation job. On the other side, the investor who is evaluating the economic feasibility of a certain submarine cable project does not want to spent money on a fully-fledged marine survey just to provide data for a bidding process.

In any case, the project planner should perform a comprehensive Desk Top Study (DTS). An amazing large amount of data can be retrieved from public sources, marine authorities, and commercial vendors. Bathymetric charts, seafloor temperatures, marine soil maps, currents, weather, waves, tides, hydrological data, other seabed users, and much more can be compiled without leaving the office [1]. The exact positions in an agreed geodetic reference system are important when compiling the data from different sources. The World Geodetic System 84 (WGDS84) is a standard in the field. The DTS also should evaluate the need for permits both for the subsequent on-site marine route survey and for the cable installation itself. Neglecting the DTS will most probably lead to increased survey and installation costs. A complete DTS is a good base plate for a bidding process except for very complicated projects.

Marine route survey means to leave the office and go on-site. Commercial survey companies would perform one or more survey runs with dedicated survey vessels to provide any wanted data along the suggested cable route. For simpler cases, vessels of opportunity (VOO) are mobilized with the equipment needed for the particular job. Either case, the quality of the used equipment and the skills of the crew are pivotal for the outcome of the effort. It is insufficient to merely identify the existence of a rock outcrop or a shipwreck, but it is essential to determine its exact position. Inaccuracies from the surface positioning system (often DGPS) and the underwater detection system may add up to intolerable values leaving the seafloor topographic data less useful for the cable installation.

The marine route survey run with the survey vessel can be amended by the sampling of seafloor material. Also, a visit on the landing site for the submarine cable is considered mandatory.

## 6.2 Bathymetry

The marine equivalent of a topographic land map is a bathymetric map of the sea bottom, indicating the depth at any point. The cable design engineer needs to know the water depth in order to determine the mechanical properties of the submarine cable. Slopes can be dangerous terrains, as cables might have a tendency to skid down the slope, or as soil movements may destruct the cables.

Bathymetric data have been used for navigation for more than 3000 years. They can be obtained from various sources with changing quality and accuracy. Only national or government hydrographical offices (HO) provide official charts and their latest updates. Commercial sea charts are available from different vendors, mostly as electronic charts (EC). In the best case, these products are a digitalized form of the official HO charts. Only a few EC systems meet the requirements of the IMO "Safety of Life at Sea" (SOLAS) code. In spite of their fancy appearance, electronic charts are in no way more accurate than the paper charts they often are compiled from. For the desk-top-study, HO charts or EC products derived from HO data can be a nice start.

The project planner should be aware of the accuracy limitation of bathymetric data. Depth soundings have been made for centuries, and one should thoroughly check the source of bathymetric data. As of 2002, over 50% of the depth information found on US charts were based on hydrographic surveys conducted before 1940. But coastlines might have changed since. Construction activities can alter the coastline considerably within a few years, in particular in harbour areas and close to densely populated areas. Also, natural erosion/aggregation has the potential to reshape coastlines and the bathymetry. Seafloors in or close to tidal waters or other water currents are particularly affected by fast changes of the seafloor bathymetry.

All bathymetry refers to a water surface reference level called chart datum (CD). Since 2005, the North Sea states adopt the "Lowest Astronomical Tide" (LAT) as CD. This is obviously lower than the mean sea level. Germany, Spain, UK, and France have adopted LAT as CD for the North Sea. In Belgium, parts of the Netherlands and Denmark (W of Skagen), the mean Low Water Springs Level is used as CD. In the Baltic Sea and Denmark E of Skagen, the CD is usually defined as the mean sea level [2]. The chart datum may have changed since the date of measurements, either due to administrative/legal changes or due to real changes out on the seafloor.

The differences on CD in relation to the on-shore levels matter little in the open sea. However, in the coastal and splash zone the tidal and level conditions can affect the cable laying operations critically.

Early chart makers could only sound the depth at distinct points. They may have interpolated between the sounded spots, allowing for the negligence of ridges and outcrops in-between [3]. Between the spotted and charted soundings, there may exist natural ridges, outcrops, shipwrecks, and other disturbing features.

A very useful source for desktop-studies in North-American waters is the service provided by the National Oceanic and Atmospheric Administration (NOAA) at http://www.nauticalcharts.noaa.gov/mcd/OnLineViewer.html. An annoying detail, however, is that the depth information of the provided charts sometimes are given in feet, sometimes in fathoms.[1] Fathom-data can be affected with a truncation error of up to 1 m. Correct bathymetric data are important both for the design of the cable and for the choice of suitable laying vessels with respect to their draught. Mistaking the sounding units can have expensive consequences.

Today, sonar-based systems are routinely used to provide a high-resolution bathymetric contour of the cable route. The method can be compared with radar, but uses high frequency sound pulses with frequencies up to 500 KHz instead of electromagnetic pulses. The systems transmit sound "pings" and analyse the return signal that has bounced off the seafloor or objects on the seafloor.

Several bathymetry methods are industry standard and can be used for the marine route survey. A multi-beam echo-sounder (MBES) with high frequency has a high resolution, relatively small transducers, and can thus be mounted on VOO mobilized

[1] 1 fath equals 6 ft. Conversion information is given in the Useful Tables chapter.

for the task. This advantage has to be paid for with a short range. Long range full ocean depth systems require large transducer arrays underneath dedicated survey vessels (=high day-rate). For the rather shallow waters of most submarine power cable projects the high-frequency systems might be good enough. In any case, the specification for the subcontracted marine survey should specify accuracy and coverage data. The accuracy is about 0.5–1% of the water depth in the beam centre (vertically down) and somewhat less at the beam edges. In deeper water the beam spread can monitor a wider corridor. Often, the cable corridor can be surveyed in a single pass, but in very shallow waters only a narrow corridor can be charted in one pass. With MBES, major obstacles in the cable route such as shipwrecks, large boulders, etc. can be identified. Large power cables and pipelines can be seen if the water is not too deep (cf. Fig. 6.1).

The measurement sensors of side-scan sonar are located in a towfish, a torpedo-shaped device pulled through the water at mostly constant height above the seafloor. It sends sonar detecting signals down and side-way and can provide an excellent imagery of the seafloor and all objects on it. The towfish is not influenced by the survey vessel roll and pitch. The detection ability of seafloor features is depending on the operating range, the vessel speed over ground, and the ping rate. According to IHO specifications, three pings per object are required in order to detect the object. Side-scan systems can also provide some information on the seafloor texture and boulder fields (cf. Fig. 6.2).

An insufficiently detailed survey may have serious consequences. A trenching plough was pulled over a seafloor where the maximum slope angles had been calculated to be 5°. However, the real slopes were much steeper. The plough was pulled over sandwaves with larger steepness, eventually rolling over on a 17° slope, and the plough was lost [5].

During the route survey for the new Wolfe Island cable in the St. Lawrence River both boulders, debris and intake pipes for a nearby plant have been discovered. Furthermore, an existing power cable was found 60 m away from its supposed position [6]. The findings helped to readjust the route for the new cable to avoid hazards.

**Fig. 6.1** Survey vessel charting the bathymetry by sonar (Courtesy of National Institute of Water & Atmospheric Research Ltd, New Zealand)

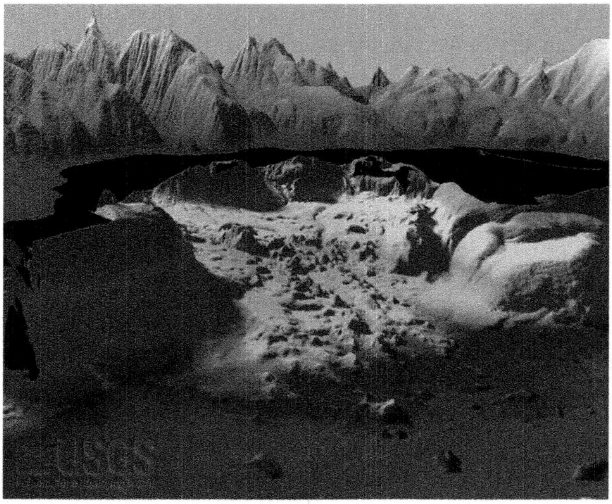


**Fig. 6.2** 3D-processed bathymetric pictures from Lake Tahoe, US. The black area is the unsurveyed coastal area. Vertical exaggeration is ×2. The debris tongue is 7.5 km wide and 9 km long. The large blocks within the debris tongue are up to 20 m high [4]

Some of the worst enemies of submarine cables are free spans where the cable hangs between outcrops, peaks, boulders and ridges.

## 6.3 Sub-bottom Profiling

If the submarine cable is to be protected, the knowledge of the seafloor bathymetry is not enough. Especially for cable burying it is necessary to know what is hidden below the sea bottom. The character, hardness, and thickness of various sub-bottom layers are decisive for the choice of the cable protection method, as well as for the assumption on the thermal parameters of the soil. Sub-bottom profiling can provide useful information on the depth and character of the strata existing under the seafloor.

The sub-bottom profiling unit is vessel based (shallow waters) or towfish based (deep waters) and is able to investigate the sub-seafloor profile vertically under the unit. Chirp-profilers produce a frequency-modulated signal providing both large soil penetration and depth resolution [1].

The sub-bottom profiling should be performed as comprehensive as possible. The sub-bottom morphology can change significantly even when the seafloor surface is not changing. Insufficient knowledge about the sub-seafloor characteristics may cause serious problems during burying of cables. All of a sudden, the cable trenching equipment might encounter a bedrock layer only a foot under the nice-looking sandy seafloor. This situation would lead to a costly variation of the cable protecting operation. Another example of insufficient survey depth: a sub-bottom profile had been performed down to the design burial depth of 2 m in a large offshore wind farm power cable project showing no problems. It turned out that the seafloor bathymetry had changed since survey. The cable plough hooked eventually into a pipe, which had been slightly under the –2 m level, such invisible for the inappropriate survey.

## 6.4 Visual Inspection

In complicated waters a survey performed from the sea surface may be too insensitive to reveal all underwater features. In order to achieve a complete picture on the subsea conditions before cable laying, manned submarine boats have been employed, e.g. during the survey for the Messina Strait (Italy-Sicily) cable system [7]. Today, similar operations can be performed rather easily by Remote-Operated Vehicles (ROV) or Autonomous Underwater Vehicles (AUV). Both types are able to carry camera systems down to the laying depth for a detailed inspection of critical route sections.

A site visit to the landing points is a must. The experienced cable installation engineer would gather valuable information on soil conditions, best locations for winches, trucks, supply, storage lots, etc. Interviews with people knowing the site under different weather, tidal, and seasonal conditions would complement the data collected in the DTS. All this helps to decide on the best possible installation technology and to foresee potential hazards.

## 6.5 Soil Sampling

Sometimes, the sub-bottom profiling has no sufficient resolution to provide data for a burial assessment survey [8]. Soil sampling can deliver additional data from selected locations to support decisions on burial tools. The hardness and character of the seafloor in different sub-seafloor depths can be investigated by taking “cores”. In contrast to remote survey techniques, coring delivers tangible samples for further laboratory evaluation (Fig. 6.3).

Beside the hardness/strength, also the thermal resistivity $\rho_T$ of the seafloor can be measured on cored samples. The thermal resistivity $\rho_T$ in the cable route is an important parameter for the design of the power cable. Although $\rho_T$ does not vary very much for many known subsea soil types, there can be spots of unusual soil with different characteristics. Especially soil types with high content of organic contents may have a high $\rho_T$ value leading to a poor cable cooling [9]. It can save a lot of money to perform a comprehensive determination of $\rho_T$ en route by soil sampling. Good knowledge on $\rho_T$ en route enables the cable design engineer to fine-tune the conductor cross-section and save money.

Soil sampling may also reveal the existence of chemical contaminations in the seafloor. Waste or ammunition dumping, or aggregation of contaminated sediments in river estuaries may have caused contaminations. The contaminations can be released during the cable installation.

The hardness of the seafloor is an important factor for the specification of the burial method.

The costs of sampling along the cable route depend on the number of sample points, i.e. the required distance between sampling points. The results of the desktop study and possibly from the sub-bottom profiling provide valuable inputs for the determination of the necessary sampling frequency. If the sub-bottom profiling

**Fig. 6.3** Core sampler

survey indicates long stretches of uniform seafloor the sample points can be 2–20 km apart. The more the seafloor properties vary along the cable route the closer the sample points must be located. Usually the survey and installation companies have the knowledge to make recommendations.

## 6.6 Soil and Water Temperatures

The variation of subsea soil and water temperatures over the year is one of the first data the cable design engineer asks for. For many waters, comprehensive data sets are available, taking annual variations and many years' statistics into account. For the submarine cable engineer the water temperature at sea bottom level is interesting, which can differ significantly from the water surface temperature.

In most submarine power cable projects, the thermal conditions at the landfalls are most onerous. High summer temperatures, large virtual burial depth in underbores and dike crossings, or interference with industrial activities must be listed before the cable design is settled. For power cables connecting offshore facilities, the ambient temperatures in the J-tubes and on the topside are necessary for a suitable cable design.

## References


1. Lyall G (2000). Minimum Standards for Subsea Cable Route Surveys, Underwater Magazine, November/December 2000. http://www.diveweb.com/telecom/features/novdec2000.01.htm
2. Bundesamt für Schifffahrt und Hydrographie, Germany, (2007). http://www.bsh.de/de/Meeresdaten/Vorhersagen/Gezeiten/811.jsp
3. Calder B R (2006). On the Uncertainty of Archive Hydrographic Data Sets, IEEE Journal of Oceanic Engineering, Vol. 31, No. 2, pp. 249–265, April 2006.
4. Gardner J V et al. (1998). USGS Open File Report 98-509, U.S.Geological Survey, http://tahoe.usgs.gov/bath980815.html
5. Allan P G (2001). Hydrographic Information and the Submarine Cable Industry. Proceedings of Hydro 2001, Norwich, UK, March 2001.
6. Canadian Seabed Research Ltd (2007). Wolfe Island Cable Route Survey. http://www.canhydro.com/projects/wolfeislandwind/project_docs/CdnSeabed_Report/WI_CdnSeaBed Report_Ch4&5.pdf
7. The CIGRE Paper (1986). Methods to Prevent Mechanical Damage to Submarine Cables, Presented by Cigré Working group 21 as Session Paper 21-12 at the 1986 Cigré Session, Paris, France.
8. Mole P et al. (1997). Cable Protection – Solutions Through New Installation and Burial Approaches. Proceedings of the 3th International Convention on Undersea Communications, SubOptic 1997.
9. Abu-Hamdeh N H et al. (2000). Soil Thermal Conductivity: Effects of Density, Moisture, Salt Concentration, and Organic Matter, Soil Science Society of America Journal, Vol. 64, pp. 1285–1290.

# Chapter 7
# Installation and Protection of Submarine Power Cables

## Contents



## 7.1 Installation

The installation of submarine cables has developed substantially in many respects during the past 20 years. Cable manufacturers are able to deliver very long lengths of submarine power cables, up to 160 km in one piece. Powerful cable laying vessels with more than 6000 t turntables can operate in heavy sea-states using satellite-based navigation systems and modern propulsion systems. A lot of experience from the oil & gas industry can be used also for the installation of submarine power cables. Also, the booming industry of submarine fibre-cables fuelled the technical progress in submarine cable laying. Remote-operated-vehicles (ROV) have developed amazing skills today, which nobody could imagine only two decades ago. Also, the seafloor

T. Worzyk, *Submarine Power Cables*, Power Systems,
DOI 10.1007/978-3-642-01270-9_7, 

survey can be performed so much easier than it could 20 years ago, today providing a close-up look onto the seafloor and, thus, avoidance of any adverse condition for the cable.

Still, the installation of submarine power cables is not an easy game. Meticulous planning is necessary, taking into account the properties of the cables, the characteristics of the cable route, and the abilities of the installation machinery. The circumstances differ so much from case to case that planning of costs and schedules must provide room for slippage, even if a good knowledge of all prerequisites exists. A successful submarine cable installation requires a carefully selected and integrated assembly of vessels, crews, and auxiliary equipment. An inappropriate cable-laying vessel, undersized equipment on board, or an inexperienced crew have devastated more than one cable laying campaign. Furthermore, a suitable determined installation management must be included. Clear and powerful authorisations for all parties on board should be agreed on.

### *7.1.1 Cable Laying Vessels*

The cable-laying vessel (CLV) is the heart of each cable laying campaign. CLVs are available at all sizes and with all kind of equipment. Worldwide there are a few fully equipped and highly specialised CLVs for large power cables, such as C/S "Skagerrak" and C/S "Guilo Verne" with payloads exceeding 6000 t of cable (Fig. 7.1).

These high-capacity CLVs have a respectable day rate, particularly in times of high demand. There is also a number of cable laying vessels with very large tonnage

**Fig. 7.1** C/S "Skagerrak" DP2 vessel with over 6000 t turntable, suitable for all kinds of power cable installation and repair

built for the installation of telecom cables. Most of these vessels are not suited for the installation of power cables as the cable handling equipment onboard is designed for light-weight small-diameter telecom cables. Some of these vessels can be used for power cable installation after considerable refurbishments.

Apart from dedicated CLVs many other vessel types can be temporarily converted for cable laying purposes. Barges and supply vessels can be equipped with suitable gear for cable laying tasks. Barges do not have own propulsion but rely on tug boats for transfer travel. During cable laying, they can be towed by tugs and/or anchors that are carefully placed along the cable route. Barges for this purpose come in all sizes and shapes.

Factors to be considered when selecting a cable laying platform (vessel or barge) are load carrying capability, manoeuvrability properties, deck space for cable handling equipment and jointing shack, crew quarter size, etc. Good sea-keeping properties, i.e. stability of the vessel in wind and waves, are essential for operations in the open sea. The bollard pull is important when a cable plough is to be used.

The barge in Fig. 7.2 has no own propulsion and must be towed to the destination. During cable laying she is pulled by anchor moorings and winches. The following features are visible: Towing chains at the left, central turntable for heavy cables, anchor winches in each corner, blue accommodation containers in three floors, white A-frame in the aft for launching rigid joints and other structures.

The barge in Fig. 7.3 is able to sit on the seafloor during low tides. The cable is coiled into a fixed cable tank. Another, rather unusual barge with a 1700 tonnes-turntable and a 6 m-laying wheel is the “H P Lading” (Fig. 7.4 ). She is the chopped-off foreship of the once-proud 1930 Exxon tanker “Esso Köbenhavn” after this was broken up in 1963. She has accommodation for 25 people.

**Fig. 7.2** Large cable laying barge (Courtesy of Oceanteam Power and Umbilicals, UK)

**Fig. 7.3** Flat bottom barge mobilised for cable laying in tidal flats (Courtesy of Oceanteam Power and Umbilicals, UK)

**Fig. 7.4** Cable laying barge "H.P. Lading". Tow barge with turntable for up-and-down spooling on the conical hub (Courtesy of NKT Cables A/S)

A new fleet of high-capacity purpose-built CLVs is under construction. Figure 7.5 shows one of the new CLVs. Important features are:

- Turntable with 6000–7000 t cable capacity
- Space aft of the turntable for a jointing house and cable tensioners
- Large wheel or chute on the aft of the vessel for deploying the cable
- High deck strength to add more cable laying facilities or temporary turntables
- Crane over the cable track to handle rigid joints
- Helipad for easy crew transfer.

**Fig. 7.5** Latest generation cable laying vessel (under construction). (Courtesy of Oceanteam Power and Umbilicals, UK)

*Load capacity*. Each laying campaign should transport as much cable as possible to reduce the number of costly and sometimes risky joints at sea. The largest CLVs for power cables have a turntable with an outer diameter of about 30 m and a load capacity exceeding 6000 MT. The load capacity is marginally dependant on the seawater temperature and salinity, i.e. the location and the time of year. Depending on the cable design, volume or weight can be limiting factors when determining the load capacity.

*Turntables*. The turntable (or carousel) has a vertical axis and is able to store even the largest of power cables, which cannot, due to their torsional stiffness, be stored in a fixed tank by coiling. Most turntables are being loaded in horizontal layers starting from the bottom layer. According to another "school", turntables have a conical inner hub and the cable is wound onto this hub in concentric layers up and down this hub under constant tension. The CLV "H P Lading" and some new vessels are built according to this design. This method requires a continuous cable tension during loading and laying. If the continuous tension is interrupted, this method can cause disorder in the cable turns as they slide down from the center.

Some vessels are or can be equipped with two independent turntables. With this equipment, two cables can be laid simultaneously. If the CLV is equipped with two spaced laying wheels[1] at the stern or bow, the two cables can be laid at a constant distance to each other. Preferentially, the cables are laid closely, bundled to each other. The bundling is accomplished as the cable pair leaves the CLV by means of suitable tape wrapping or cable straps. This is the preferred method for a pair of HVDC cables in shallow waters (up to 250 m water depth.). In the latter case, the pair of cables can be buried later in a single operation.

[1]The laying wheel represents all type of laying gear such as wheels, chutes etc.

Turntables can be divided into an inner and an outer partition. Two cables can be loaded into these partitions independently, either one after the other, or into both partitions at the same time (Fig. 7.6). This method requires a uniform-lay-direction cable armoring. Basically, both cables can be wound into their respective partition of the turntable independently by using individual feeding speeds out from the factory. A better method is, though, to wind up the cable into the outer partition and to coil down the other cable into the inner partition over a hanging coil spreader. Since the turntable is rotating to wind up the first cable, the second cable goes through a semi-coiling where the cable is twisted less than one time per turn. During laying, the process is reversed and the cables can be laid at the same common speed onto the seafloor without residual torsion. The pair of cables can be laid and bundled so that they can be buried in a single run later on. If the cables are not bundled, they might diverge on their way to the seafloor and thus jeopardize the subsequent burial.

For smaller submarine cable projects or repair jobs, small turntables can be mounted temporarily on the decks of barges or supply vessels.

*Fixed cable tanks*. Circular or oblong holds for cables are common for telecom cables and can be used for power cables, which are not torsion-stiff. The cable is loaded via a coiling spreader that is hung or fixed in considerable height above the tank. While loading, it is important to maintain a minimum inner coiling diameter specified by the cable manufacturer. Supports must be provided to prevent the cable from sliding down from the growing cable stack. A gang of coilers walking on top of the coiled stack must put the cable in place. For the sake of their safety, the coil

**Fig. 7.6** Turntable with inner and outer partition for simultaneous laying of two cables. Here, a submarine power cable (outer partition) and a fibre-optic cable (inner partition) are stored on the turntable (Courtesy of ABB, Sweden)

stack should be fenced as it grows in height. The laying is the reverse operation of the loading, and the cable should land on the seafloor without internal torsion. Fixed tanks can be used for cables of moderate diameter with uniform-lay-direction armoring. Single-core cables with single wire armoring are good candidates for the use of fixed tankes. Medium-voltage three-phase submarine cable can also be loaded in a similar technique in oblong holds in vessels.

*Cable drums.* For many submarine cable projects with short cable length, it is not necessary to employ a dedicated CLV. Submarine cables connecting offshore wind turbines have a length of 400–800 m. They can be installed from barges or supply vessels equipped with cable drum pay-offs and suitable breaking and tensioning engines (Fig. 7.7). Extremely large drums with horizontal axis can be found on some pipeline laying vessels and, basically, may be used for cables.

*Positioning.* Every cable-laying vessel must be able to position itself very accurately on the prescribed cable corridor and move in desired directions without loosing positioning control. Even slight deviations from the planned position and heading can adventure the health of the cable or the accuracy of cable laying. The cable may end up in a subsea position different from the anticipated one. Maybe the cable goes down beside a pre-dredged trench, outside a licensed corridor, or in a hazardous seafloor area.

*Anchoring systems.* The classical method to keep the CLV in position is the use of anchoring systems. The CLV, often a propulsionless barge, is manoeuvred between the holds of numerous anchors fanning out in all directions. The anchors are handled by independent AHT (anchor handling tugs). Cable laying barges employ four to

**Fig. 7.7** Laying with cable drums on a barge

eight anchors. They are placed some hundred or even thousand metres away from the barge into the water and connected to winches onboard the barge. The barge controls its position, speed, and heading by operating the winches. In very shallow waters it can be advantageous to use flat barges and anchoring methods and tugs because large CLVs often have large draught and protruding propulsion devices. The anchoring method is risky in waters where other cables or pipelines exist. These lines are hooked easily by the numerous anchors, resulting in costly repair jobs. Also, the anchors must not damage the cable already laid. The anchoring method is a tedious and time-consuming method but it avoids the high day-rates of a self-propelled CLV.

*Dynamic positioning systems (DP)*. These systems keep the vessel on a determined position by means of high-sophisticated navigation systems and various ship propulsion devices. DP systems can move the vessel along a pre-determined course over ground. DP systems can also keep the vessel on station with any desired heading. Within certain limits, this is possible even when wind, waves, or currents try to carry the vessel off course. Conventional screw propellers provide forward thrust and, in combination with rudders, also directional control to a certain degree. A number of other propulsion systems are available to provide directional control, even without rudders and at low or no speed. Tunnel thrusters in bow and stern can push the vessel sideway and turn it on the spot. Azimuth thrusters are screw propellers, suspended in pivotable pods under the vessel (Fig. 7.8). They can turn 360° in most cases, and provide excellent manoeuvrability at any speed. Two independent azimuth thrusters can keep the vessel on station or turn it into the required

**Fig. 7.8** Azimuth pods (Courtesy of ABB Oy, Marine, Finland)

heading. The pod propellers can be powered mechanically from an engine in the hull (Schottel drive), or with an electric motor in the pod powered from a generator in the vessel hull. Some vessels have retractable azimuths to improve the ability to navigate shallow waters.

Voith-Schneider propulsions consist of rotating vertical blades mounted under the vessel. They can generate thrust in all directions. Azimuth drives and Voith-Schneider propellers provide excellent manoeuvrability and can move the vessel in all directions (back/forth, sideways), which can be necessary during complicated cable laying or repair operations. In contrast to conventional propulsion, they do not need a minimum speed to maintain direction and control. The vessel is able to lay the cable along a well-defined corridor and still to keep the heading into the weather to reduce roll. For stationary jobs such as jointing or ROV operations the vessel is able to keep station and direction also under rough conditions.

DP systems use different navigation systems such as GPS-supported navigation systems, taught wire systems, and acoustic beacons. Most of modern vessels have GPS based systems. They are classified into DP-classes DP0 through DP3 depending on the redundancy of subsystems according to an International Maritime Organisation (IMO) classification. Redundancy requirements include not only propulsion and navigation systems but also features such as independent power generation systems and other auxiliary systems separated by firebreaks. A higher DP class on the vessel increases the safety of the cable operation. The DP class should be carefully selected with respect to the requirements and risk profile of the project. Sometimes, insurance companies require the use of vessels with a defined DP class. Many large submarine cable projects employ DP2 vessels.

*Jointing house*. The jointing of submarine power cables can only be performed in specially equipped jointing houses. They must be designed according to the specifications of the jointing company and need to have sufficient size. For a recent submarine power cable project, the on-board jointing house had a size of $4 \times 17$ m. The jointing house must be equipped with electric power supply, air condition, and air dryer. It must be fixed sea-worthily on an appropriate place in the cableway on deck. It is of advantage if the jointing house is splitable so it can be removed from the cable after jointing. For safety reasons, the jointing house should have redundant direct radio communication to the vessel bridge, and an alarm bell commanded from the bridge.

*Cable tensioners*. Also called linear machines, they are necessary for any moving of cables on board and for applying tension during laying. Many linear machines comprise pairs of wheels. The wheel pairs can be opened and closed with controlled pressure to grip the cable. Individual opening is necessary when bulky joints must be deployed. Wheel drive is most often hydraulical. The linear machine in Fig. 7.9 has eight wheel pairs and is installed close to the laying wheel at the vessel stern. The first end of the cable is being passed through the linear machine, which is in pulling mode. When the cable is being deployed into the water, the linear machine can be operated in braking mode. Many machines can be operated in speed- or tension-controlled mode. Linear machines are also available with belts instead of wheels, so called caterpillars. A good friction between the wheels and the cable surface is

**Fig. 7.9** Cable tensioner with eight wheel pairs taking grip of the leading cable end

necessary without undue pressure on the cable. Not unlike car tyres, the wheels on the linear machine need to have a sufficient grip for efficient braking/pulling power, and the cable surface must not be too slippery. Some cable layers advocate the use of sand between cable and wheels when the grip slips.

The need for sufficient power in the linear machines must not be underestimated. The cooperation of linear machines, turntables and the vessel must be orchestrated carefully.

*Emergency cutter*. As described below in the weather chapter, a situation may arise when the cable must be cut in emergency situations. The cutting must be performed rapidly. The preferred solution is a hydraulic cutting head close to the aft wheel with remote control (cf. Fig. 7.10). A simpler device is a movable disc cutter with petrol or electric motor to be operated by a man. The emergency cutting device should be able to cut the cable within 60–90s.

*Cableways, rollers, laying and pick-up arms, chutes, and laying wheels* are other devices necessary for the cable installation. For each specific installation job, the ensemble of machinery must be selected to meet the requirements and allow for a safe and successful operation. Especially the specified bending radii under defined tension values should be taken into account. It is wise not to try to save money when specifying the equipment for an installation job. Many operations suffered painful delays because of weak cable rollers that have been blown away under heavy loads, or because of too weak linear engines.

It has been discussed to design aft laying wheels or chutes with dynamic heave compensation, in order to facilitate cable installation in higher sea-states. Another

**Fig. 7.10** Hydraulic emergency cable cutter

means to reduce the vertical movement of the laying wheel is to arrange it midships, in order to reduce the effects of heave and pitch.

*ROV equipment*. Many activities in cable laying operations are possible only with the help of ROV (remote operated vehicle, Fig. 7.11). The ROV is a submersible powered tool carrier, which can be equipped with different manipulators and tools. Cameras are valuable to inspect difficult route sections, boulders, outcrops, or man-made leftovers such as wrecks or containers. Also, in easy waters, the ROV can provide valuable data for the as-built documentation. The ROV can handle submarine pick-up operations, inspections of possible damages, and close-ups of questionable areas. ROV are highly complex systems and are available in all sizes. The vessel should have sufficient space for a complete ROV system, comprising the vehicle and its parking space, launch crane, umbilical drum, container for control room and supplies, accommodation for the ROV operating crew, etc.

*Helicopter landing pad*. For the crew change or transport of specialists like jointers or fault locators a helicopter transfer can be an economic or even the only possible option.

### *7.1.2 Other Vessels*

Apart from the CLV, other vessels are often needed for a submarine cable installation. In strong winds or currents, the CLV may need the assistance of one or more tugs to keep position. During landing operations, a fleet of smaller vessels

**Fig. 7.11** ROV with grip manipulators and cameras

is often busy with the handling of pull wires, anchors, floating devices, and the cable. Anchor handling vessels are needed when the cable installation is performed by a propulsionless CLV. For complicated routes, the use of a survey ship during installation can be essential.

For post-lay cable protection a mother vessel for the trenching or jetting equipment is necessary. Before completed protection a fleet of guard vessels is useful to chase away fishing boats and protect the cable from unauthorized access.

Installation jobs far away from the coast require accomodation for the crew. If the CLV does not have sufficient accommodation a separate hotel vessel must be employed. Accommodations must be approved by authorities and insurance companies.

Manned submersible vessels have been used occasionally for cable related jobs [1]. Today, highly sophisticated ROV′'s can perform almost any underwater job without risk.

A 123 kV three-phase cable was installed in 2006 between the two Thai islands of Koh Samui and Kha Nom over a distance of 24 km. Five vessels were participating:

- one laying barge (85×24 m) equipped with 6-point-mooring system, cable tank, LCE, diving spread, jetting spread.
- two AHT (anchor handling tugs)
- one accomodation vessel
- one shallow water jetting sledge support vessel.

### *7.1.3 Loading and Logistics*

Submarine cables with short length can be handled and transported on standard or oversize drums. Most standard and all oversized drums require flatbed trailers making the onshore transport expensive.

When the submarine cable factory is located next to a harbour, the cable drums can be transported to the destination by a cargo vessel. Oversized drums can be shipped in this way avoiding the troublesome and expensive heavy road transports. Some cable manufactures ship long cables in coils laying on a group of rail cars [2], (cf. Fig. 7.12).

The length of in-field cables in offshore wind parks is often in the range of 400–800 m. There are two concepts for the supply of the cables:

1. Supply of pre-cut lengths on drums. Loading and installation require but rather simple machinery and barges with low day rates. However, the correct lengths must already be known in the factory. The requirement of relatively large safety margins in length results in large amounts of scrap cable during installation. Also, the empty drums must be either disposed of or returned, which can be a

**Fig. 7.12** Transport of submarine cable on railway cars (Coutesy of NKT Cables A/S, Denmark)

costly activity. Another disadvantage can be a possible client requirement for individual factory acceptance tests on each cable drum.
2. The submarine cable can be shipped in a large coil or on a turntable. On-site, the lengths are cut from the coil as needed during laying. With this concept, less scrap cable and no drums need to be disposed of. However, more complicated equipment is needed both in the factory and on-site.

For the loading of a long cable length, the CLV is mooring directly at the cable factory to take up the cable in one piece. From the factory turntable, the cable runs over cable rollers to the CLV over a laying arm onto the ship turntable. Depending on cable size and equipment capacity, the loading speed is 3–20 m/min. The complicated loading operation requires the synchronised action of turntables and linear machines continuously over days or even weeks. Failures in synchronisation of speed can render kinks, bends, or buckling of the cable. This kind of damages may require costly and time-consuming repairs. Any possible measure to reduce the risk of loss of synchronisation must be considered prior to loading.

Other methods to load long cable lengths manufactured in factories at some distance from the harbour have been used in few cases: Cable roller tracks have been used by a Swedish company, and a British company shipped many drum lengths to a jointing ground at the harbour site. Both methods are tedious and risky.

The loop time for the CLV (load, transfer to site, manufacturing of connecting joint to previously laid cable, laying of the new cable, return to factory harbour) is most often much longer than the laying time. When the installation site is far away from the cable factory, the speed of the transportation vessel can be important. CLVs are relatively slow. Separate cargo vessels can speed up the cable transport, but this scheme requires a transfer of cable from the cargo vessel to the CLV at a port close to the installation site.

The ends of the cable must always be sealed properly to avoid intrusion of water or even humidity. The use of inadequate capping methods or equipment may result in water intrusion, which may require a cut-away of a considerably long length of cable. In the worst case, the cut-away of water-damaged cables requires the use of extra spare cable and extra joints. The end caps are manufactured according to the manufacturer's specifications. Shrink-on end caps often are not appropriate.

The water and humidity barriers in the submarine cable must not be damaged during transport, laying and burying. Too small bending radii, large side wall pressures, shear forces, etc. can destruct metallic laminates and metallic sheathes.

### *7.1.4 Laying of Submarine Power Cables*

Submarine cables on drums, e.g. for in-field cables, can be installed from barges with simpler navigation equipment. A drum pay-off and a linear machine with brake are necessary. The cable runs over a chute or a laying wheel into the water. The chute or wheel must have sufficiently large a diameter to maintain the MBR. The barge normally has no propulsion and must be manoeuvred by tugboats or a set of anchors and winches. Stabilizing the position of a barge requires at least four anchors, one on each corner. Once the anchors are set, the barge can move by hauling in and paying out the anchor lines in an orchestrated manner. The anchors need to be relocated from time to time by specialized anchor handling vessels. To keep the vessel on station at all times, it takes therefore more than four anchors, possibly up to eight or even more. This method is risky to other cables or pipes in the area, and it has actually happened that anchors from the laying barge have damaged cables already installed in the same project. Altogether, this is a very slow method, making 1–2 km a day as a maximum. However, in some waters, the use of barges able to rest on the seafloor at low tides might be the only way to get the job done. In these waters, ROV cameras are often useless as the visibility can be very little or zero.

A linear machine is necessary also for the installation of cables from a turntable. The cable runs from the motorized turntable over a movable pickup arm with cable rollers, through the linear machine, and then over the chute or wheel. The pickup arm normally has the shape of a gooseneck. In the gap between the turntable and the gooseneck, the cable can hang loose. This slack is useful to absorb possible temporary speed differences between turntable and linear engine. The gooseneck is pivotable and often also adjustable in height. Its position is moved according to the point where the cable leaves the turntable. Operators try to keep the slack always in the same curvature. This can turn out difficult, if the turntable drive works depending on load or position of the turntable, or if the linear machine has slippage or wheelspin. The cable runs further towards the laying wheel or chute before departure into the water. Close before the wheel or chute, another linear machine can be installed, working in breaking or pulling mode. For greater laying depth, a Capstan wheel may be used to amplify the cable breaking force from the linear cable engine.

When designing the installation, one of the machines is defined as the "master". This machine defines the laying speed and all other machines, including the vessel propulsion, must follow the master speed.

If the cable would be let down to the sea floor vertically, there would be a risk of looping and instability of the laying direction. The heaving, pitching, and rolling movement of the aft sheave can cause cable damages at the touch down (TD) point, as the cable might be compressed longitudinally. Instead, the cable must be positioned in a well-defined catenary line from the laying wheel to the TD by application of a certain forward tension by means of the vessel. Under these circumstances, the cable hits the sea floor in a flat angle. A fully equipped CLV has the possibility to monitor the catenary parameters. Sensors to monitor the departure angle and the lay tension provide information to calculate the other catenary parameters. Equations relating water depth, bottom tension, top tension, departure angle, and the horizontal distance between aft sheave and TD are given in the Appendix at the end of this chapter.

An inappropriate catenary line or bottom tension can be corrected by a temporary extra pay-out from the turntable, or by a temporary acceleration of the vessel speed. At critical passages of the cable route, the TD should be monitored by ROV.

The bottom tension is considered a critical parameter for the laying. Too low a bottom tension can cause the cable to build loops or to snake, especially if the cable is laid from a fixed tank, which generates twisting in the cable. Many cable laying "pilots" consider that a high bottom tension results in a cable on the seafloor carrying a highly residual tension. The residual cable tension is thought to impede later burying for protection. Figure 7.28 in the Appendix of this chapter shows the main parameters of the catenary line.

Misalignment of the various speeds (vessel, turntable, linear machines, tensioners, etc.) can cause a cable kink, loops or damages of the deck equipment. Submarine power cables that break out from their roller guides during laying may cause serious damages and hazards for the deck crew.

A power cable stored in a fixed cable tank can be paid-out using a fixed or movable pick-up arm over the tank. During loading in the factory, the cable has captured one torsional turn for each turn it was laid around the cable tank. This residual torsion is now released when the cable enters the fixed cable guide on the vessel. The rest of the laying from a fixed tank is equal to laying from a turntable.

Single-wire armored cables are not torsional-balanced. As the cable tension at the laying wheel and the bottom tension are unequal, there will be a resulting torsion in the cable as it goes down to the seafloor. At great depths especially, this may lead to cable loops. The catenary of these cables should be monitored by ROV – if possible – to detect any irregularities.

When a pair of cables is installed simultaneously, the cables can be stored in two independent turntables. A double set of cable handling equipment transports the cables to the laying wheel. Before departing, the cables are bundled together using steel straps, tape wrapping heads, or something similar. Bundled cables can be buried later in a single run. For the Cross Sound and the Estlink HVDC projects, the two HVDC cables and a fibre-optical cable were bundled and laid simultaneously.

For cables stored on inner and outer partitions of a turntable, different pay-off equipment is necessary. Most often, the outer cable has been spooled onto the turntable in the cable factory without torsion. It will be taken up as described with a movable pick-up arm, and paid out. The inner cable must reach the laying wheel with the same speed as the outer cable. Independent from the turntable speed, it will be taken up with a gooseneck and pulled by a tensioner with a speed that corresponds to the laying speed of the outer cable. However, as the turntable is actually turning to pay out the outer cable, only part of a torsional turn is released in the inner cable for each turn it is uncoiled from the turntable. For this, the method is called "semi-coiling".

There is a large difference in cable laying complexity, if a plain feature-less seafloor allows for a "simple" reel-out of the cable, or if the seafloor structure requires that the cable be laid between and around obstacles such as boulders, rocks, or outcrops. Free spans should be avoided under all circumstances. Sometimes, ROV-supported monitoring is necessary in order to direct the cable into the best possible line between the obstacles. This is also valid for pipe and cable crossings. Sometimes, special laying techniques must be developed and engineered for critical cable routes.

ROV monitoring of the TD is recommended even for other reasons:

1. An inappropriate TD angle can damage the cable seriously. Using ROV on-line monitoring can alert suitable action at an early stage.
2. Equipped with positioning device the ROV can supply very precise position data of the cable route for the as-built documentation.

#### 7.1.4.1 Laying of Cable Around a Curve

When laying the cable around a bend with too small a bending radius, the residual bottom tension tends to drag the cable laterally over the seafloor. The minimum required bending radius can be calculated as

$$R_s = L_s = \frac{FT_H}{W_s \mu} \tag{7.1}$$

where (cf. Fig. 7.13)

$F$, Safety factor (suggested: 2.0);
$T_H$, Horizontal bottom tension;
$W_s$, Cable unit weight in water;
$\mu$, lateral friction coefficient cable-soil.

After passing an obstacle, a straight line of length $L_s$ should be laid before the curve is initiated. This is to avoid the cable being dragged against the obstacle as a result of lateral displacement. For cables with smooth extruded outer sheath, the following friction coefficients $\mu$ may be assumed: clay 0.2, sand 0.6, and gravel 0.8 [3]. Slightly higher values can be used for cables with yarn serving.

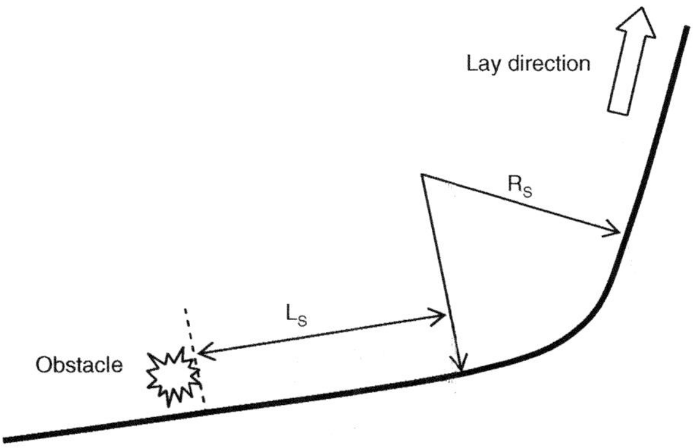


**Fig. 7.13** Laying the cable in a curve

## 7.1.5 *Landing of Submarine Cables*

The landing of submarine power cables sometimes requires the most engineering, the most equipment and often the most time of all efforts of the cable project. The majority of all submarine cable projects have at least one landing point. Regularly, the laying operation is started here. A number of methods for the cable landing have been used, but new projects may require new methods. The method to be used depends much on the shore conditions, equipment abilities and authority requirements. In cable installation jargon the coastline is called beach even if nobody wants to spend a single day of vacation there.

Sometimes it is possible to establish an open trench through the beach. The entrance point can be stabilised by cofferdams. The open trench goes from the sea entrance to the most appropriate land site for the joint pit between the submarine cable and the connecting land cable, called beach joint. The beach joint can be located inside the real sand beach, behind a dike, or as far away as some kilometres from the beach (e.g. the Baltic Cable project). The trench routing should be as straight as possible in order to avoid undue pulling forces and curves. During pull-in, the cable is guided on roller guideways inside the trench or outside; in the latter case, the cable will be pushed into the trench afterwards. The trench profile must be engineered with respect to thermal properties. Despite the fact that dikes along the European North Sea coast are very sensitive areas and normally must not be penetrated under any conditions, the NorNed cable could be landed through the Dutch protection dike system in open trench thanks to a well prepared operation (Fig. 7.14). Only one year after the operation the dike appears in pristine conditions.

In some areas, the only acknowledged method to cross the beach zone is the use of closed pipes. The method of choice is Horizontal Directional Drilling (HDD), where a drilling station is erected on the land side of the dike or beach. The drilling station drills a hole with a pre-determined curvature under the beach area and/or dike into the open water. The drilling path can be controlled in all directions. After drilling, the hole is lined with steel or plastic pipes for the later pull-in of cables. The pipe sections must be welded without leaving any internal welding beads, which could obstruct the cable pulling through the pipe. The longest possible length for

**Fig. 7.14** Landing of a submarine power cable (NorNed) in open trench over the protection wall. In the splash zone, an open pit was created with cofferdam walls to receive the cable. The cut in the dike is visible in the distance in the *right* picture. Cable winch is in the foreground

HDD for cable applications is today in the range of 1400–1800 m. The length is limited not only by the drilling technique but also by the maximum pulling length of cables. The latter depends on:

- cable properties like weight, stiffness and maximum allowed pulling forces
- friction coefficients
- bending radii
- lubrication methods
- possibility of employing cable pushing, e.g. at river crossings.

The risk of failure increases more than linearly with the pulling length. For long pull length, the internal diameter of the pipe should be 2.5 times the cable diameter, for a short pull length, a factor of 1.5 is sufficient. Pulling two or three cables in the same pipe increases the pulling forces substantially and is recommended for very short length only (e.g. road crossings). The necessary pulling force for a straight pipe can be estimated by the following equation:

$$\text{Pulling force} = \text{cable weight} \cdot \text{friction coefficient} \tag{7.2}$$

With a pulling length of 400 m, a cable weight of 400 N/m and a friction coefficient of 0.4, the necessary pulling force is 64 kN (~6.4 MT). This is valid for a straight pipe without internal obstructions such as weld beads. The maximum pulling force can be decreased when the pipe is water-filled to obtain bouyancy for the cable. Proper choice of materials for cable sheath and pipe walls can decrease the friction coefficient. Friction coefficients for various material pairings are depending on the explicite composition of the polymeric materials. Tabulated values are of little use in practical life as the properties often are influenced by temperature and dirt on the cable surface. For engineering purposes, the friction coefficient can be set to 0.4 for unlubricated cable pulls, and 0.25 for lubricated cable pulls. Lubricating agents can be mineral slurries, gels, or bio-degradable oils. A slight bend, such as a gentle curve under a river crossing, does not change the figure very much. However, pipe bends increase the necessary pulling force substantially and should be avoided with all methods. Any bend contributes strongly to the resulting pulling force, especially if the bend is located close to the target end of the pulling route. If possible, cable pulling always starts from the difficult bends.

When all land preparations are concluded, the CLV approaches the landing spot as close as possible and stays there by means of its DP system. On shore, a winch is installed and a pull wire from the winch is transported to the CLV and connected to the cable end. The power cable can now be paid out over the laying wheel or the chute. Floating devices (air-filled bags) are being attached to the cable, while more cable is being paid out. Water currents e.g. due to tidal changes and wind may interfere strongly with this operation. The curvature in the cable in Fig. 7.15 was caused by tidal longshore currents.

**Fig. 7.15** Floating of a submarine power cable during landing (NorNed). Crew members wear safety harness

Auxiliary boats may be necessary to keep the cable in position, while the shore winch hauls in the pull wire and the power cable to the beach and further to its destination. When the cable end has reached the destination on shore, it can be anchored there and the air-bags can be removed starting from the beach. The cable is now sinking down to its predetermined position. The maximum pulling length on land is much depending on the route. The Baltic Cable, a 450 kV HVDC cable, was pulled for about 5 km from the beach onto land, using a large number of small synchronised cable pulling machines along the route. The above-mentioned methods can be applied for the starting end of the cable. After landing the cable with the above-mentioned method, the CLV can start laying the cable. At the destination beach, different methods must be employed. There, the CLV arrives with the end piece of the submarine cable, which must be brought onto land. The CLV would move towards the beach as close as possible. From now on, the cable would be equipped with floating devices such as air cushions. Then the CLV would turn into a heading in parallel to the beach or even back. It is important that the floating devices are attached to the cable before the CLV starts changing its direction. The CLV now pays out the remainder of the cable, or at least as much as is necessary to reach the beach jointing pit. Since many cables are produced with spare or excess length there will be a remainder of cable onboard the vessel. A pulling eye is attached to the properly cut tailing end of the cable when it leaves the CLV. The floating cable can now be pulled from a shore pulling winch, when necessary assisted by working vessels. Cable layers, pray to your God that it is long enough!

Should weather conditions get worse, threatening the finalization of the destination beach approach, the CLV may reel out the remaining cable quickly in parallel to the shore line and put the cable end on the seafloor with suitable capping. In a later operation, it would recover the cable and complete the installation within a suitable weather window.

The landing of submarine cables to offshore platforms (or offshore wind-turbines, OWT) is somewhat different. Offshore structures with fixed foundations usually carry J-tubes that reach from the seafloor up to the top-side of the platform. The J-tubes are made from steel or polymeric materials. Above the top opening of the J-tube, there is a winch to be used for cable pull-in. The lower opening of the J-tube is bellmouth-shaped and points away from the platform or OWT foundation. The opening may be at, under, or above seafloor level. To install the start end of the submarine power cable, the CLV would approach the J-tube until it reaches a distance corresponding to the water depth or more. A pulling wire is sent from the topside winch down through the J-tube, and further on to the cable end onboard the vessel. The cable is paid out from the stationary vessel, while the pulling wire is hauled in. Ideally, the cable approaches the bellmouth horizontally. The pull-in of the submarine power cable is only successful when some conditions are fulfilled:

- The bending radius of the J-tube must be sufficiently high. It is most advisable to design the J-tube with a bending radius substantially larger than the MBR of the cable in order to avoid undue friction and high tensional loads on the topside winch
- The J-tube interior must be clear of obstacles and welding beads. The J-tube should be camera-inspected before cable pull-in. The internal diameter of the J-tube should be 2.5 times the cable diameter
- The winch is sufficiently strong
- An ROV or a diver shall be provided to camera-monitor the bellmouth.

The landing of the second end of the cable at the next OWT or platform can be performed in a similar way. The difference to the fore-end is that the tail of the cable must be taken off the CLV and parked somewhere, before it can be pulled into the J-tube. The "parking-lot" should be as close to the receiving J-tube as possible to reduce the length of pulled cable. The parked cable can rest on the seafloor when the water depth is in the range of 10–30 m and the seafloor is free of boulders, outcrops, and other obstacles. The parked cable can then be pulled into the J-tube from its "parking-lot" on the seafloor. In greater depths, or when movement over the seafloor is being hindered, the cable end must be parked with floating devices.

### 7.1.6 Jointing of Submarine Power Cables

Although modern cable laying vessels can store and handle enormous lengths of cable, it is sometimes unavoidable to joint cables on open sea because the length of the cable route cannot be covered with a single shipload of cable. Other applications of submarine cable joints include the jointing of different cable designs along the route. The design of submarine cable joints is treated in Chap. 4.

The assembly of submarine power cable joints on open sea is a challenge that requires meticulous planning, suitable and reliable equipment, and a highly specialised and well-trained crew. With these preconditions, and a sufficiently long period of favorable weather, the joint can be manufactured and deployed safely. The manufacturing of the joint inside the jointing shack is described in Chap. 4. Here, the positioning of cables before jointing and deployment of the completed joint is to be discussed. The deployment of a manufactured joint on the seafloor can be a complicated operation, as the two jointed cables must be handled simultaneously with the joint. No overbending or overtensioning must occur, nor must the cable arrangement get stuck in A-frames or other structures on board. The cable laying circumstances decide which type of sea joint can be used.

Power cable joints must be prepared in purpose-built jointing houses onboard the CLV. A simple tent might be acceptable for the jointing of medium-voltage cables in calm waters. The jointing of high-voltage submarine cables requires jointing houses with air-conditioning, air-drying facilities, hoisting, and cable handling equipment. Outside the jointing house, additional space might be required for the air-conditioning units, storage, and waste boxes.

When laying long lengths of submarine cables in subsequent campaigns it is sometimes not possible to joint the next length immediately to the previous length. It is possible to sink a cable end into the water for an indefinite time and later recovery if the cable end is capped carefully according to the manufacturer's instructions. The use of inadequate methods or equipment may result in water intrusion, which requires a cut-away of a considerably long piece of cable. In the worst case, the cut-away of water-damaged cables requires the use of extra spare cable and extra joints.

#### 7.1.6.1 In-Line Joints

In the simplest laying scheme, the CLV is laying cable **A** first. The end of Cable A is provided with a cable seal with pulling eye, a ground wire, and a hooking arrangement before it is lowered to the seafloor. The hooking arrangement should be ROV friendly for later recovery. Buoyancy links at the end of the ground wire can be useful, should sediment movements bury the ground wire.

After the first laying campaign, the vessel returns to port to fetch the next length, Cable B. The CLV returns to the cable route at the end of cable **A**. The ground wire of cable **A** is hooked with an ROV or a grapnel and pulled onboard over the laying wheel to a jointing shack onboard the vessel. There, the second cable length, which is still onboard the vessel, is jointed to the first cable. If the joint is a flexible joint (cf. Chap. 4), the completed joint can now be transferred through the ordinary laying equipment of the vessel, and passes over the stern laying wheel down into the water. The vessel is keeping a suitable cable tension and layback in order to preserve the joint from undue bending at the touchdown point. The jointed new

Cable B is following, and the laying operation can be continued. This arrangement requires that the complete second cable runs through the jointing shack unless the shack is a split design and can be removed after jointing. If the joint design is rigid, it cannot be pulled through the ordinary cable route and over the stern wheel. Instead, the rigid joint body must be hauled over carefully with cranes. At the same time, sharp bends of the cables must be avoided where they are connected to the rigid joint body. A carefully orchestrated operation of deck cranes, vessel positioning, and winches is necessary to deploy the joint, bring Cable B fully into the cable laying machinery, and lay the joint correctly onto the seafloor. It should be avoided that the joint position on the seaflor is affected by sharp bends, boulder fields, and free spans.

In-line joints are comparatively easy to handle, particularly if flexible joints can be used. The cables lie straight on the seafloor and can later be buried in a single uninterrupted run (precaution is needed when a rigid joint body is too bulky for the trenching equipment). However, the method requires a long period of good weather for the jointing operation and the subsequent laying of Cable B.

A different approach is to lay Cable B without jointing it to cable A in the first place. Cable A and B will later be connected by a so-called hairpin joint.

### 7.1.6.2 After-Installation Joints

Sometimes it is necessary to lay down both Cables A and B onto the seafloor without jointing. A possible reason might be a short weather window not allowing for both jointing and subsequent laying in one go. Also, it can be less costly to use the high day-rate CLV only for laying, and use a cheaper vessel for the jointing job. The post-lay jointing can be done in different configurations. One post-lay cable configuration is depicted in Fig. 7.16. Both Cable ends A and B are lying on the seafloor with straight overlap, each hopefully equipped with seal, pulling eye, ground wire, and ROV-friendly hooking devices.

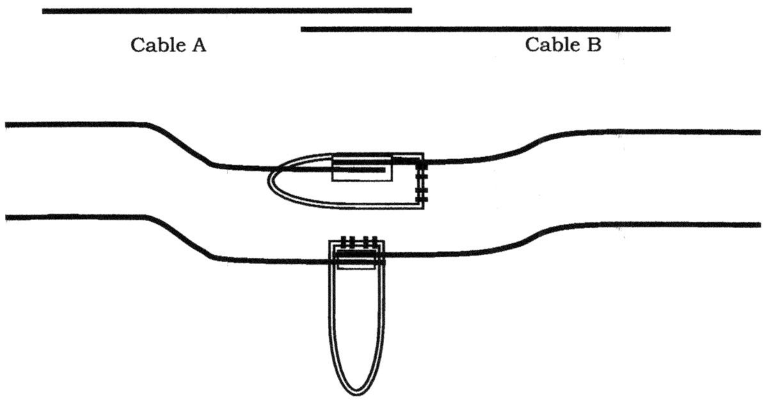


**Fig. 7.16** Two cables laid with overlap, and positioned in a jointing shack (symbolized by a rectangle on the vessel) parallel to the bulkward or across the aft deck

For the final jointing, the two cable ends can be pulled up to a jointing shack arranged along the deckside and the bulkward. The ground wires of the cable ends are hooked by ROV or grapnel, pulled up and guided to the jointing shack over wheels, chutes, and suitable cable guides. The length of the overlap in Fig. 7.16 must be two times the water depth, plus the length of the desired catenary lines during jointing, plus 100–150 m depending on the length of the cable guides onboard. It can be prudent to have an extra length to provide cable for a second try if the first joint must be redone.

Another similar set-up is to have the jointing shack transversely across the aft deck of the jointing vessel (cf. lower part of Fig. 7.16). The Cables A and B would be pulled up over chutes or wheels protruding laterally from the vessel, and guided into the jointing shack using winches, linear machines, or other suitable equipment. The ends are positioned inside the jointing shack with overlap according to the joint manufacturer's instructions. The cables must be secured firmly outside the jointing shack so they cannot move under any conditions.

The vessel must keep station during the subsequent jointing operation. If the DP system cannot keep the vessel on station in difficult weather tug boats can assist to stabilize the position of the CLV. After jointing the jointing shack must be removed or opened so that the jointed cables can be deployed into the water.

Another configuration for a post-lay jointing is shown in Fig. 7.17.

The cable ends are laid departing from the straight cable route. The distance between the two cables in the spur route may range between 5 and 20 m. The length of the spur route must be sufficient in order to keep the cable in a stable position on the seafloor even if the CLV keeps pulling in order to maintain a proper catenary line during the jointing. The minimum length $L_S$ for stability is given in Eq. 7.1 in this chapter. Further allowance must be made for 100–150 m on board of the CLV. If provisions shall be made for a second try of joint manufacture, the spur cable

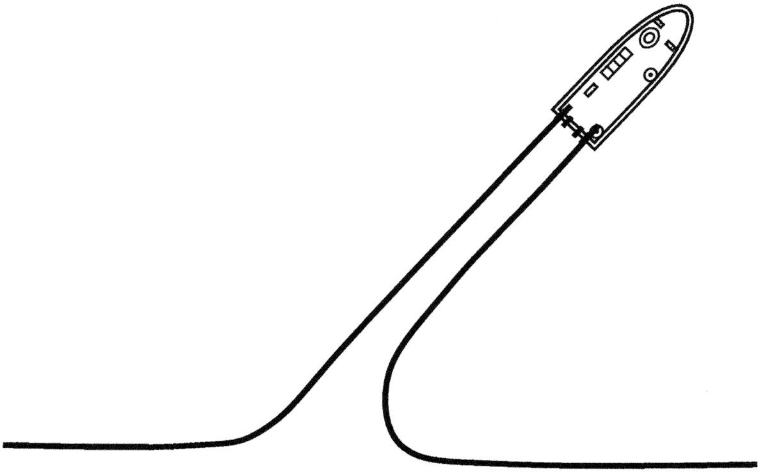

**Fig. 7.17** Two cables laid for a hairpin joint

length must be even longer. The spur cable route may be laid in a sharper angle to either side in order to provide an advantageous CLV heading into the waves for the jointing operation to come.

For the preparation of the hairpin joint, both cable ends are pulled up to the vessel aft deck simultaneously, while the CLV travels backwards along the spur route. One cable is being pulled up over the ordinary stern wheel, the other over a provisional wheel or chute. Well onboard the vessel, both cable ends are guided to the jointing shack using winches, linear machines, or other suitable equipment. The ends are positioned inside the jointing shack with overlap according to the joint manufacturer's instructions. The cables must be secured firmly outside the jointing shack in a way that they cannot move under any conditions. The CLV must keep station during the subsequent jointing operation. After jointing the jointing shack must be removed or opened so that the jointed cables can be moved. The jointed cables **A** and **B** form a loop or hairpin-shaped open bend which now must be deployed into the water. The deployment of the joint is possibly the trickiest part of the operation. The excess length of the two jointed cables must be organized to form gentle curves on the seafloor. A welded steel arc ("quadrant") is used to support the cable in the loop and provide the correct bending radius (cf. Fig. 7.18, and the description of an example repair operation in Chap. 7). During the deployment, the vessel moves along the spur route to maintain suitable cable tension.

In the trade, the hairpin that is deployed on the seafloor is also known as "loop", "omega", or "final bight". More nicknames are probably being used.

Suitable vessel equipment provided, a hairpin joint can be installed with both flexible and rigid cable joints. The advantage of hairpin joints is that it can be pro-

**Fig. 7.18** The quadrant, a steel support frame, goes into the water (Courtesy of ABB, Sweden)

duced independently of the cable laying operation, maybe on a later occasion with more favorable weather. The hairpin joint can also be manufactured using a different vessel with appropriate equipment for the operation. This vessel can have a lower dayrate than a top-of the-line CLV. A hairpin joint provides no straight cable corridor and post-lay trenching must take special measures to protect the spur route and final bight.

## *7.1.7 Weather*

Weather is the cable crew's worst enemy. Countless are the hours during which an armada of cable laying vessels and working boats has been rolling idly in waves and wind for hours, days, or even weeks until useful weather appeared from the skies. Weather can demolish any working plan and any schedule. Neptune is still invincible if it were with the fanciest technology. Waiting on weather ("WoW") is a very costly way of doing nothing.

Unfortunately, it is almost impossible to list weather limits for a safe laying operation. Neither wind nor waves per se constitute obstacles for cable laying. What matters is only the movement of the laying wheel when the cable is hanging down from it. This movement is caused by waves, which in turn are caused by winds. Each step in this causal chain involves statistical relations and local parameters making things difficult.

### 7.1.7.1 Winds

Wind generates waves. The wind distribution can be very different for different places and is subject to strong seasonal variations, which may create weather windows of different length for the cable laying. The wind speed is given in the classical Beaufort scale or other speed units such as km/h, m/s or knots (1 kn = 1852 m/h). Relations are listed in Table 7.1 together with the characteristics of related sea-states.

### 7.1.7.2 Wave Properties

Waves originate from wind blowing over a water surface. Fresh wind waves usually have a short time period and can be rather steep. The longer the length of the route where wind can attack the water ("fetch") the higher the waves can build up. In parts of the North Sea, westerly winds have a longer fetch and build up much higher waves than easterly winds. In the Baltic Sea, parts of the Mediterranean, and other marginal seas the fetch is limited and the wave height is often smaller than with similar wind forces over the Ocean.

The size of waves that wind of a given strength can generate depends also on the wind duration and the water depth. Empirical formulae for the calculation of wave characteristics from wind speed are given in [4] for the Baltic and North Sea. Table 7.2 summarizes some of the results with the condition of constant wind speed in constant direction over the entire area.

**Table 7.1** Wind speed and sea-states

| | Wind speed | | | | |
|---|---|---|---|---|---|
| Beaufort | Kn | km/h | m/s | Description | Sea conditions |
| 0 | 0 | 0 | 0–0.2 | Calm | Flat |
| 1 | 1–3 | 1–6 | 0.3–1.5 | Light air | Ripples without crests |
| 2 | 4–6 | 7–11 | 1.6–3.3 | Light breeze | Small wavelets. Crests of glassy appearance, not breaking |
| 3 | 7–10 | 12–19 | 3.4–5.4 | Gentle breeze | Large wavelets. Crests begin to break; scattered whitecaps |
| 4 | 11–15 | 20–29 | 5.5–7.9 | Moderate breeze | Small waves |
| 5 | 16–21 | 30–39 | 8.0–10.7 | Fresh breeze | Moderate longer waves. Some foam and spray |
| 6 | 22–27 | 40–50 | 10.8–13.8 | Strong breeze | Large waves with foam crests and some spray |
| 7 | 28–33 | 51–62 | 13.9–17.1 | Near gale/ Moderate gale | Sea heaps up and foam begins to streak |
| 8 | 34–40 | 63–75 | 17.2–20.7 | Fresh gale | Moderately high waves with breaking crests forming spindrift. Streaks of foam |
| 9 | 41–47 | 76–87 | 20.8–24.4 | Strong gale | High waves with dense foam. Wave crests start to roll over. Considerable spray |
| 10 | 48–55 | 88–102 | 24.5–28.4 | Whole gale/storm | Very high waves. The sea surface is white and there is considerable tumbling. Visibility is reduced |
| 11 | 56–63 | 103–119 | 28.5–32.6 | Violent storm | Exceptionally high waves |
| 12 | 64–80 | 120 | 32.7–40.8 | Hurricane | Huge waves. Air filled with foam and spray. Sea completely white with driving spray. Visibility greatly reduced |

The values given here serve for an approximate overview and must not be taken for engineering purposes. It is highly recommended to use wave spectra for the actual location of a proposed submarine cable project.

Waves generated by winds can travel a far way over the oceans. Travelling over hundreds of miles, the waves become less steep, having longer periods, but can

**Table 7.2** Wave height at different wind speeds

| | Wind speed | | | |
|---|---|---|---|---|
| | 20 kn<br>10.3 m/s<br>37 km/h<br>5 Bft | 30 kn<br>15.4 m/s<br>56 km/h<br>7 Bft | 40 kn<br>20.6 m/s<br>74 km/h<br>9 Bft | 46 kn<br>23.7 m/s<br>85 km/h<br>10 Bft |
| **Baltic Sea** | | | | |
| Significant wave height $H_S$, m | 1.6 | 3.3 | 5.3 | 6.3 |
| Highest of 100 waves $H^{1/100}$, m | 2.4 | 5.0 | 8.0 | 9.5 |
| **North Sea and Mediterranean (west of 21°E)** | | | | |
| Significant wave height $H_S$, m | 1.9 | 4.8 | 8.6 | 10.3 |
| Highest of 100 waves $H^{1/100}$, m | 2.9 | 7.2 | 13.0 | 16.0 |

conserve much of the wave height. These waves, which are generated far away are called "swell", and may prevail for days after they have been generated. Travelling waves can be reflected from coastlines. In some spots, travelling waves from different origins can interfere and build up to larger wave heights than those from each contributing wave train. Refraction and interference of waves can build up complicated wave patterns particularly between coastal islands and the mainland, or in waters where sand banks can impact wave train directions [4]. For this reason, the waves at a certain location are composed of waves with different characteristics and height. Wave direction can be different from and even opposite to wind direction. Waves can be existent in the absence of wind, as a result of earlier winds/storms.

Each sea-state comprises a spectrum of waves with different wave height. The wave height distribution is often described as a Raleigh distribution. In many sea-state forecasts the term "significant wave height" is used. The significant wave height $H_s$ is defined as the average wave height of the highest third of all waves [5, 6]. In practice, one would measure the height of 10,000 subsequent waves and take the average of the highest 3333 waves to obtain $H_s$. The definition of $H_s$ is independent of the statistical distribution of the wave heights. If the Rayleigh distribution is used the probability of various wave heights can be related to $H_s$:

$H^{1/10}$ (one wave out of ten in the upper third) $= 1.27 \cdot H_s$
$H^{1/100}$ (one wave out of hundred in the upper third) $= 1.67 \cdot H_s$
$H^{1/1000}$ (one wave out of thousand in the upper third) $= 1.86 \cdot H_s$
$H^{max}$ (Max wave height expected) $= 2 \cdot H_s$

Figure 7.19 shows Raleigh-distributed wave heights for a significant wave height of 4.4 m (indicated by a vertical line).

Other researchers prefer other statistical distributions (e.g. Gauss distribution) to represent the wave spectrum. This disagreement reflects the fact that the wave spectrum differs between locations. A comprehensive treatment of wave theories and a related bibliography can be found in [7]. Freak waves (also called "monster

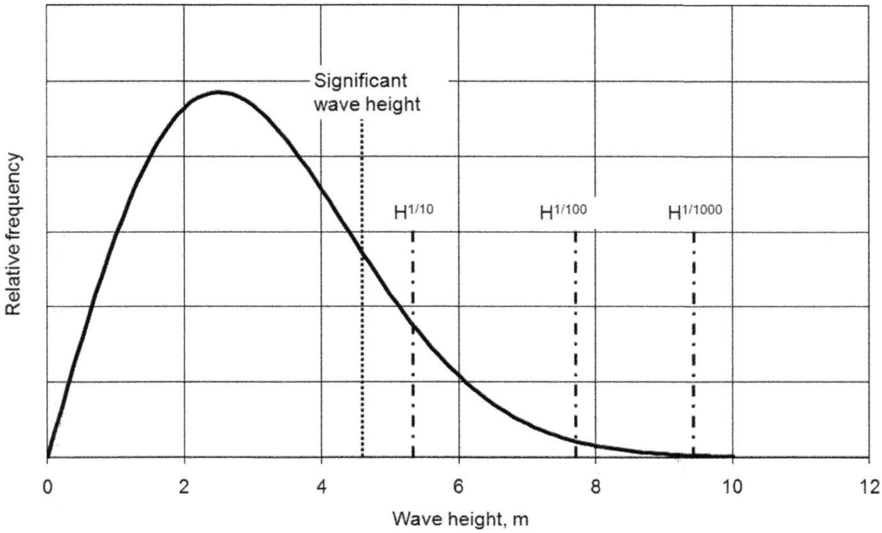


**Fig. 7.19** Wave height of a Rayleigh-distributed wave spectrum. Significant wave height is indicated by a *vertical line* at 4.4 m

waves" or "rogue waves") are not part of these statistics. Today the existence of freak waves is widely acknowledged [4, 8].

For the cable installation engineer, it is important to know that the maximum wave height can be roughly twice the significant wave height given in sea-state forecasts, and that waves with $2{\cdot}H_S$ are expected to occur a few times in each eight-hour interval. As the accuracy of sea-state forecasts decreases for longer forecast horizons, the cable installation engineer should consider increasing the safety margins for cable installation operations with long duration.

#### 7.1.7.3 Vessel Movement

The next question is how different vessels react to a given wave spectrum. The response is not only depending on vessel size but also on its design and equipment. For each CLV a RAO (Response Amplitude Operator) can be established that takes into account an individual vessel's size, mass distribution and other parameters reflecting the seakeeping properties of the vessel.

The movement of a cable laying vessel (CLV) has six degrees of freedom (Fig. 7.20):

The RAO (Response Amplitude Operator) for the vessel in question translates wave characteristics to vessel movements. In particular, the heading of the waves with respect to the vessel is an important factor to determine the vessel heave, pitch, and roll from the wave characteristics. For laying operations, pitching and heaving have the largest influence on the vertical movement of the laying wheel. Still, rolling cannot be totally neglected if the aft laying wheel is off the vessel centre line.

| | |
|---|---|
| Lateral movements | Heave (up and down)<br>Surge (back and forth)<br>Sway (right-to-left) |
| Rotational movements | Roll (rotation around the longitudinal axis)<br>Pitch (rotation around the beam axis)<br>Yaw (rotation around the vertical axis |

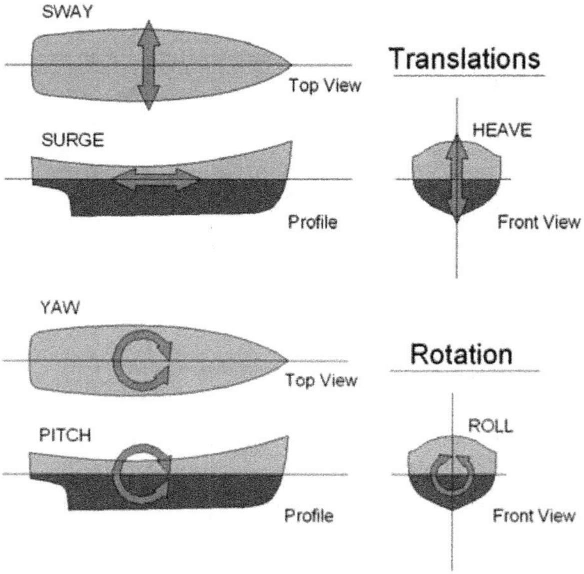


**Fig. 7.20** Six basic movements of a ship. Source: Wikipedia, http://en.wikipedia.org/wiki/Ship_motions

Figure 7.21 shows the RAO factor for a specific cable-laying vessel in the 6000 t payload class. The period time of the (sinusoïdal) wave is represented on the $x$-axis. The $y$-axis is the RAO factor for heave. The factor is given in m/m, and describes the relations between the resulting heave amplitude to the causing wave amplitude. A RAO heave factor of 0.6 means that the heave of the vessel is 0.6 times the wave height. Seven curves are given for different wave headings related to the vessel. 180° means head-on waves, 90° are beam waves. Some interesting observations can be deducted from Fig. 7.21:

- The vessel response on waves is very much depending on the wave direction and the wave period time.
- Waves with short period time usually do not cause large heave since the wave length is smaller than the vessel length.
- Beam waves can cause large heave even at relatively short period time.
- In very short and very long waves the wave direction is less critical for the heave.

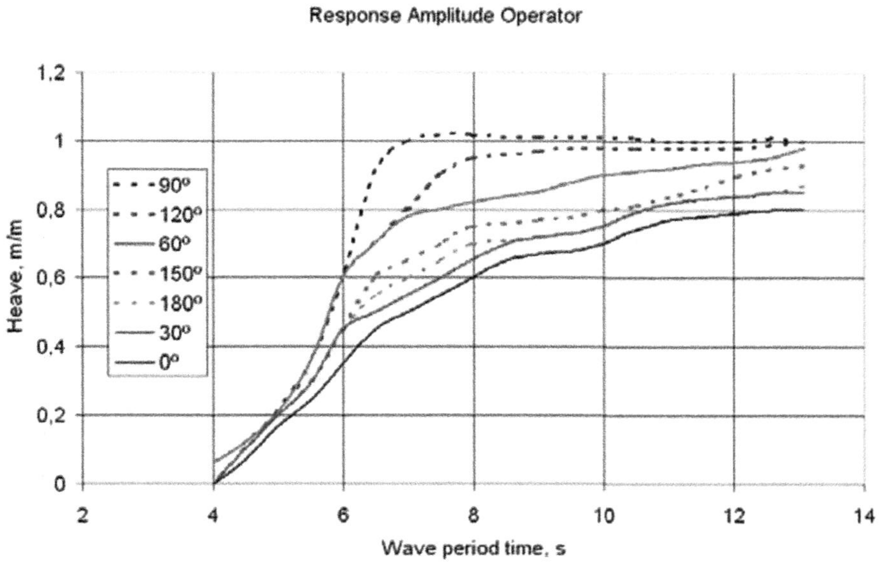


**Fig. 7.21** RAO for heave for a 6000 t payload CLV

During cable laying, the vessel heading is dictated by the cable route. If the cable route is S–N and the prevailing winds are NW, the CLV receives the waves from a quarterly direction, which often results in the most onerous movements.

The cable that is hanging down from the laying wheel in a catenary line is subjected to a number of forces. Let us first consider a simple case when the cable hangs down almost vertically. Assume that the cable has a weight $w$ per meter in water (that is the weight in air minus the cable's buoyancy). The water depth is $D$. The vessel stands still. The weight of the hanging cable is now

$$F = D \cdot w \tag{7.3}$$

Now, we assume that the laying wheel is moving up and down in response to waves. To make it easy, we assume that the vessel floats on the waves like a wine cork, i.e. the laying wheel movement is identical to the wave movement. The movement has a sinusoidal shape. The movement amplitude (measured peak to peak) is $h$, and the movement period (time between subsequent upside peaks) is $T$. The laying wheel movement can now be described as:

$$y = h/2 \cdot \sin(2\pi/T \cdot t) \tag{7.4}$$

The largest acceleration of such a movement can be found by twofold derivation with respect to time and amounts to:

$$a_{\max} = h/2 \cdot (2\pi/T)^2 \tag{7.5}$$

As an example, a vertical laying wheel movement with 4 m peak-to-peak and a 6 s period time reaches a maximum acceleration of about 2.2 m/s$^2$. The maximum force on the hanging cable is now

$$F_{\max} = F + m \cdot a_{\max} \tag{7.6}$$

where $m$ is the mass of the hanging cable. Since the force due to acceleration is an inertia phenomenon, the cable mass must be used for the calculation rather than the weight in water or air.

Actually, this value can be larger still. Especially high waves with short period can cause substantial acceleration values. In addition, waves can interfere, which leads to sudden vertical movements. Stowing guidelines published by the International Maritime Organization (IMO) and International Labor Organization (ILO) of 02.05.1997 state that vertical acceleration of up to 1 g (=9.81 ms$^{-2}$) can be observed at the aft and bow of a vessel [9]. However, these high acceleration values are encountered on vessels under speed into head waves, which results in shorter wave periods. The speed of CLVs is rather low (mostly under 1 kn) so the contribution of vessel speed can mostly be neglected. Still it is prudent engineering to consider an appreciable vertical acceleration when estimating the cable tension.

During cable laying with a departure angle of 25°–30°, the suspended cable length is considerably higher than the water depth. The entire cable must be accelerated following the vertical movements of the laying wheel. The cable portion far away from the wheel, close to the TD, is moving up and down with, in comparison to the laying wheel, lower amplitude. The necessary force to accelerate the cable must be vertical. The cable can only transmit forces along its own direction. For this reason, large tensions along the cable have to be transmitted to create the vertical forces for the cable acceleration close to the TD.

The tensional forces on the cable are results of waves with a statistically distributed wave height. Even if the significant wave height of the weather forecast seems to be harmless, only a few waves of exceptional amplitude can cause tensions big enough to damage the cable.

#### 7.1.7.4 Other Impacts of Wind and Waves

Vessel movements can introduce adverse effects on cable laying beyond the tensional force on the cable:

*Crew work ability*. Many seamen suffer from seasickness. In heavy weather, 10–30% of the crew might be sick and unable to work [10]. The healthy part of the crew might be unable to work for safety reasons.

*Danger to the crew*. The cable jointing crew inside the jointing shack can be subjected to danger from moving equipment. All equipment in the jointing house must be lashed to the walls or the floor, and working can be very difficult when the jointers need one hand to secure themselves all the time. The power cable must be anchored firmly on board to keep it in place. The greatest danger comes from the cable in the jointing shack. In very strong heaves, the cable hanging in the water is exposed to forces so strong that it might be overtensioned. The grip of cable tensioners might slip or the cable securing devices on board might break.[2] The situation is out of control, when the heavy cable starts now moving unfastened in the jointing shack and over deck areas. This is seriously dangerous for the crew, the equipment, and the vessel. It calls for immediate crew clearance of the area and possibly emergency cut of the cable in order to save health or life of the deck crew. Emergency cutting is a safety measure and causes normally large economic losses in form of costly repair operations and schedule slippage. Should the Master or the charterer of the vessel, in order to save consequential costs, put the crew's safety at risk, he can be subject to criminal prosecution.

*Loss of position.* The CLV keeps station by means of thrust engines. In heavy weather, the thruster capability can be overwhelmed by side winds, strong currents and/or waves. The vessel is driven from its determined position. The cable is still in its hanging position and may act as an anchor chain. It can be subject to additional tensional strength and also to other forces, when the direction of drift is not along the cable route. This has happened in a number of cable installation projects. When the CLV is been driven back over the cable, there is a risk of twisting and kinking of the cable.

### *7.1.8 Organisation*

In 1964, a report on a submarine cable laying operation declared: "Captain and crew of the boats shall be cooperative with the cable laying work and, if possible, having experience in cable handling" [11]. The insight behind this statement was not new. However, more than one project manager wished that everybody had read this report. Today, the installation of a major submarine link is a considerable industrial construction project. Only with meticulous planning, appropriate resource allocation and thorough contingency preparedness the project can succeed.

A number of different companies can be engaged in a cable installation such as the following parties:

- The client, represented by the client rep on board
- Nautical crew of the vessel from Master to maid
- Cable laying and handling crew onboard, perhaps from another company
- Cable manufacturer representative

---

[2]Cable tensioners must never be used alone to secure the cable during jointing.

- Survey company responsible for staying in the cable corridor, making fresh in-situ surveys, and preparing the as-built documentation
- Owners and operators for the ROV equipment, guard vessels, auxiliary vessels
- Insurance companies.

These companies can be contracted independently by the client, or can have a subcontractor hierarchy, or anything in-between. In many projects, the different tasks of the cable laying operation are organised by a single contractor who is using subcontractors. The single contractor is the only interface to the client, and the coordination responsibility is put on the single contractor. This is a convenient structure for the client as he has only one contractor to talk to. It is fair, when the single contractor is claiming extra compensation for his efforts of interface coordination and increased risk exposure.

In a different business plan, the client goes shopping for all different supplies and services in separate contracts. This gives the client the opportunity to procure the best offer for each of the contracts and to avoid paying coordination surcharges to a general contractor. In this case the interface coordination is an effort to be performed by the client (or an external company hired for that purpose).

In any case, there must be clearly defined responsibilities, order flow and mandates, and communication paths between the participating parties.

It can be of advantage for the client to hire contractors who have been working together successfully in previous projects.

Coordination and planning meetings involving the parties should be conducted repeatedly and long before the operation starts. Flaws in the project organisation or/and the installation equipment can be detected during these personal meetings. People meet, people accept the other's roll, and people know whom to talk to. The importance of a mutual understanding of the role and competence of each party can't be overestimated. Costs for early coordination meetings pay back manyfold.

During the cable installation, daily operation meetings are conducted involving representatives for all relevant parties. Apart from the HSE (Health, Safety, Environment) topics, the progress of the last 24 h and the planned work for the next 24 h are to be discussed then and there. In separate tool-box-meetings, the details of certain work activities are organized with the relevant crew.

## 7.2 Protection of Submarine Power Cables

Submarine cables are precious assets and need to be protected from external hazards. The 1986 Cigré study [14] presents a comprehensive compilation of submarine cable faults and protection methods. The protection constitutes a considerable part of the total investment.

There are four principle steps of submarine cable protection:

1. selection of a suitable cable route
2. design of suitable cable armoring
3. protection on the seafloor e.g. by burying
4. active after-installation protection.

It is widely accepted that a well-designed cable protection enhances the reliability and, hence, the availability of the cable system. The operational costs for repair and maintenance can be reduced by a proper protection.

### 7.2.1 Selection of a Suitable Cable Route

After a desktop study a provisional cable route can be selected. As far as possible, hazardous areas should be avoided, such as:

- Shipping lanes, anchorages, harbour entrances
- Fishing grounds
- Boulder fields, outcrops, submarine canyons and steep slopes
- Wrecks, ammunition-dumping grounds, debris
- Areas with strong water currents.

Shipping lanes should be avoided because the presence of heavy ship traffic can impose restrictions on the installation operations. Also, the risk of anchor damages is higher where ships travel. If shipping lanes have to be crossed somewhere, it is advisable to do this at right angle to keep the interference low. Anchorages, ammunition-dumping grounds, and military exercise areas are strict no-go's for submarine cable installation.

Fishing grounds exercise a direct threat to the cable's health. Not only fishing nets but also ancillary gear such as mooring points can do harm to the cable.

Landing points for the cable link must be selected carefully. Shorelines have very different characters and are sometimes stable in their position, sometimes very volatile and prone to changes. The shoreline and the adjacent bathymetry may change rapidly due to tidal current, longshore currents, wave patterns, seasonal storms, and more. Even if many coastal communities try to preserve the coastline as it is by various costly measures, there are voices advocating the acceptance of natural changes of the shorelines.

Shorelines may also be disturbed by man-made interference such as breakwaters, harbour entrances, beach protection, etc. When selecting a cable route, also future near-shore developments should be taken into account.

Electric stray currents from on-shore activities (electric trains, welding, etc.) have reportedly caused cable damages. Therefore, the route selection should also include possible effects from onshore activities.

Many other obstacles may be hidden in the water or even under the seafloor. The following list has a few of these, and many more different things have been found by the survey crews: ship wrecks, abandoned cars and trucks, lost containers, dumped construction materials, active or defunct subsea installations, military installations, garbage dumps, intake or outflow pipes, etc. Most often it is not possible to avoid the crossing of cables and pipelines but one should select suitable crossing points at a distance from shore.

OWP installations define a safety zone of 500 m for their own protection but it is strongly recommended to keep away from OWP at least 2000 m, unless the cable in question runs to that OWP. This is to enable possible future repair activities, which may require some space outside the cable corridor. Also, future OWP must be taken into account. The laws of the High Seas do not grant a safety zone around the submarine cable.

Great care should be devoted to the selection of the best seafloor topology and morphology for the cable route. Not only boulder fields, outcrops, submarine canyons, and steep slopes threaten the cable or call for trouble during installation. A smooth sandy seabed might seem suitable for laying and installation, but there might be a ridge of rock just some feet below, making the subsequent trenching very difficult.

Sand waves on the seafloor might constitute considerable risks for the cable life. Sand waves can move on unpredictable paths creating unexpected patterns through the years. As a result, the cable can be buried much deeper than previously expected, or the cable can ly exposed and without protection in the valleys between the sand waves. In worst cases, free spans may be generated, or the exposed cable may be sand-blasted by the sand-loaded current.

It can be more economic to run the cable in a detour rather than spend money for a higher degree of other protection in a dangerous area.

Route selection also includes fine adjustment during cable laying where ROV can monitor the detailed cable route and helps laying the cable between rocks rather than over them. In a Norwegian project, the cable was allowed to follow the "valleys" between seafloor elevations, rather than a "blind" straight laying (Fig. 7.22). Actually, hiding the cable between natural elevations might give it a good protection without burying it, but it requires meticulous control of the cable touchdown during laying.

### *7.2.2 Design of a Suitable Cable Armoring*

The armoring of submarine cables (power or telecom) must be designed to meet the tensional forces during laying and the protection requirements during the lifetime of the cable. The telecom submarine cable industry has developed an armoring

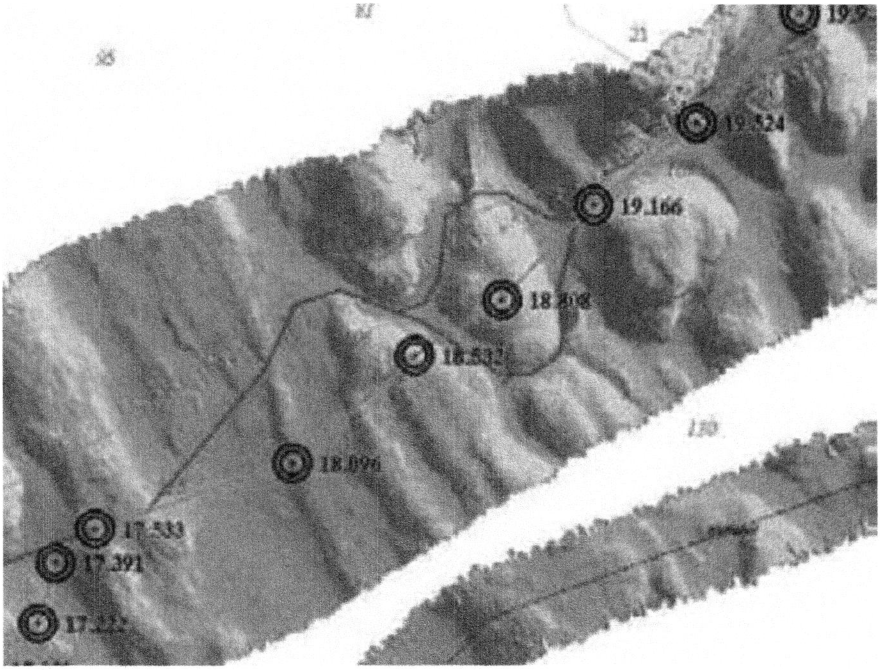


**Fig. 7.22** The initially planned route (*straight line*) would have implied several ridge/valley crossings with risk for free spans. Detailed route survey enabled to lay the cable along the *curved* route avoiding strong undulations [12] (Courtesy of Submarine Telecom Forum)

philosophy according to Table 7.3. According to this philosophy, the shallow water cables are protected with heavy armoring against external violence, while the deep-sea cables have a lighter armoring because there is little threat from human activities down there.

Submarine power cable projects in the 1950–1980s have often followed a similar concept providing a heavier armoring for the beach section as external dangers are

**Table 7.3** Types of cable armoring for submarine telecom cables [13]

| Water depth | Armoring | Characteristics |
|---|---|---|
| < 200 m | Rock Armor (RA) | Double armor with short lay in the outer armoring layer, improved impact resistance and better flexibility to follow seafloor undulations |
| <500 m | Double Armor (DA) | Protected cable for areas with little or no burial depth |
| <1500 m | Single Armor (SA) | Used for areas with limited burial depth |

expected to be more frequent in shallow waters and in the splash zone. The open sea cable was covered with a lighter protection. The principle is very easy: more massive steel wire armoring provides a better protection against fishing gear and anchors.

As submarine power cable dimensions and weight have grown considerably during the last two decades, manufacturers are less keen to provide extra heavy armoring for the beach zone. In many cases, the armoring is designed only to meet the requirements of tensional forces during laying. As most power cables are installed in shallow to moderate depths (< 300 m), the armoring can be designed rather simple. Often, this allows for a thin-wire armoring, which is not prepared to cope with external threats. Unfortunately, the protection against damages is sometimes neglected when manufactures try to submit the lowest bid. However, there are good examples as well. The Channel Island cable between France and Jersey/Guernsey was equipped with an extra-strong steel wire "rock armor" in 2000. A double-layer counter-helical armoring provides a much better protection against external violence than a single-layer armoring. In many encounters of submarine power cables with installation gear, anchors, etc., the outer wire layer was damaged but not the inner layer.

It should be considered carefully if the circumstances of the cable route (shipping lanes, fishing grounds, etc.) or the installation method may call for additional reinforcement of the cable. Often, the manufacturer can offer armoring with higher protection level as an option.

The choice of the appropriate armoring is a difficult balance between up-front capital costs and expected (better: feared) costs of future cable damages including repair and income losses. The initial costs for extra heavy armor are known but the future costs can only be projected estimates. It should be taken into account that the armoring cannot be upgraded after a few years of project experience.

Shorter parts of the cable can be equipped with additional protection by attaching plastic shells during laying of the cable (cf. Fig. 7.23). This method is often used to strengthen the protection at cable and pipeline crossings.

### 7.2.3 External Protection

Many submarine cables have been laid unprotected on the seafloor except for the immediate beach zone until the 1980s and 1990s. Until then, it was almost taken as inexorable fate that submarine power cables are being damaged from time to time. When appropriate subsea burial equipment became available in the 1980s, the burial protection of longer cable stretches became more common. Existing cables such as the Baltic Cable were buried afterwards to upgrade the protection level. When the Cigré study on mechanical damages to submarine cables [14] was compiled in 1986, only a small fraction of submarine power cables were externally protected. The study also accounts for the fate of the Kontiskan link between Denmark and Sweden. The first cable there was laid unprotected in 1964, designed to withstand all what was known of fishing gear at that time. Fishing gear grew heavier with time, and during the early 1970s, the cable was damaged too many times by trawls. It was

**Fig. 7.23** Protective plastic shells are attached to two HVDC cables and an FO cable before the cable group leaves the vessel (Courtesy of ABB, Sweden)

decided to replace a piece of the cable by a cable with heavier armoring, and to bury the new cable piece. The new cable piece had no failures until the report was written, while the remaining unburied cable suffered another 14 failures during the same time.

Today, almost all submarine power cables have an external protection (burial or covered). This protects the cable not only against accidental damages but also against sabotage. For over 100 years, submarine cables have been the target of war-related assaults; today evil forces might try to attack unprotected submarine power supply. Having this in mind, the insurance companies would probably reward suitable protection efforts with a lower premium, or even require complete cable protection prior to issuing an insurance policy.

#### 7.2.3.1 Trenching

The most common protection method today is trenching, i.e. the burial of the cable under the seafloor.

There is a variety of different trenching methods, and new equipment is being developed as the amount of submarine infrastructure increases.

Ploughing down submarine cables has been known and practised for many decades. A underwater cable plough has a horizontal framework that can travel over the seafloor either on four sledges, one in each corner, or on wheels, or caterpillars (Fig. 7.24). In the centreline of the frame there is a plough share reaching down into the seafloor. In contrary to an agricultural plough, the cable burying plough is not designed to turn over the soil but to cut a narrow slit with as little resistance as possible. While the ploughshare is cutting this slit the cable is guided down directly behind the ploughshare or through a hollow slot inside the ploughshare. The ploughshare must have a shape and inclination such that it is not pressed upwards

**Fig. 7.24** Cable plough. The tiltable plough share to the right is now horizontal, ready for tilting down to the vertical position (Courtesy of IHC Engineering Business Ltd., UK)

over the seafloor while moving. The backward inclination of the ploughshare adds vertical forces that keep the plough in the soil [15].

Ploughs are towed after the surface vessel and can require high tow forces. In deeper water, the plough requires long heavy tow cables and the position control of the plough becomes difficult [16].

More sophisticated plough systems employ vibration or water jets to support the ploughing process. These systems require power supply from the surface. The plough depth is up to 3 m depending on the seafloor stiffness.

The ploughing has some inherent risks when the plough meets rocky soil conditions. Rocks and boulders encountering the course of the plough are sometimes able to derange the heading direction and misalign the plough with respect to the cable. This has the potential to squeeze and damage the cable. Another risk is that boulders or debris can enter the entrance bellmouth together with the cable and develop enormous side impacts on the cable.

Rocks and boulders may misalign the plough in relation to the cable that in some cases can be damaged. In soft seafloor soils (sand, silt, mud, clay) a plough can be used to open a temporary furrow into the seafloor.

Ploughing down cables is a brute force method suitable for soft to medium strong soils and shallow waters. Many ploughs are passive systems with a fairly good reliability, which makes ploughing to one of the most economic cable burial methods.

#### 7.2.3.2 Jetting Methods

Another group of burial equipment relies on water jetting action. A sword carrying a row of water nozzles is pushed down into the seafloor (Fig. 7.25). The high-pressure

**Fig. 7.25** Cable jetting device. The sword carrying water nozzles is visible to the *right*, tilted up (Courtesy of LD Travocean, France)

water flow from the nozzles fluidises the seafloor (Fig. 7.25). While the water-jetting unit is moved along the cable, the cable sinks down into the slurry. Soon after, the slurry re-solidifies. Often, there are two swords travelling on each side of the power cable. Different carrier systems are available for the water-jetting unit. The simplest carrier is a sledge that slides on the seafloor much similar to a plough. A support vessel tows the sledge. These machines often feature a combination of plough and water jetting. Submersible pumps mounted on the sledge can create a water flow of 1100 $m^3/h$ at 5.5 bar pressure. One particular sledge carries six of these powerful pumps, plus a long row of monitoring instruments.

The water-jetting unit can also be assembled into a purpose-built ROV with own propulsion. This ROV is a large truck crawling on the seafloor on wheels or caterpillars. The vehicle carries high-power water pumps to create large water flows at high pressures, the jetting swords, wheel drives, cable tracking equipment, process monitoring sensors, and positioning systems. Other water-jetting ROV are free-flying over the seafloor. An umbilical cable connects the ROV to the surface vessel in order to provide power supply and data communication. The ROV-based water-jetting system is complex and requires auxiliary systems (power supply, crane, deck space, control cubicle, etc.).

#### 7.2.3.3 Simultaneous or Post-Lay Burial?

Ploughs and water-jetting equipment can be used for simultaneous laying and burial of the cable, or for post-lay burial (PLB). A PLB operation can be performed at convenient occasion after the cable has been laid by the CLV. A smaller vessel can be hired at lower day rates for the burial operation. In a PLB operation the burial equipment (plough or water-jet of any kind) moves along the cable route, guided by joystick in an on-board control room. ROV cameras provide an overview over the situation to take measures in case not only the cable but also rocks or debris are going to enter the cable entrance of the equipment, which could lead to damages on the cable. The cable is, however, at risk in the time window between laying and protection. Despite the deployment of guard vessels patrolling the cable route to chase away fishing boats, it has happened that submarine cables waiting for burial protection have been hooked.

In a simultaneous laying and burial operation, the plough or water-jetting would follow the CLV immediately. The cable can be fed into the burial equipment directly from above, which leads to a better cable guiding and reduced risk for encounters with rocks. Also, the guiding of the burial equipment on the seafloor is fairly easy. A simultaneous laying and burial operation slows down the laying operation considerably because the burial equipment is the slowest of the entire spread. This is particularly costly when a high day-rate specialized CLV must be used for the laying operation. Also, the weather window for the slow operation must be much longer than for the laying operation without simultaneous burial. Many cable installation specialists advocate the opinion that the cable should be brought into the water as quickly as possible without the retarding effect of simultaneous trenching.

The economic situation is different when the laying is performed with a less expensive laying barge. The progress of a laying barge is much slower than that of a CLV due to the time-consuming anchor handling procedures of the laying barge. Therefore, the simultaneous use of trenching gear would not have the same retarding effect as with a self-propelled CLV.

Laying barges are often used to install cables in shallow waters or tidal flats, where large CLV cannot operate. In these cases, the operation of the trenching equipment can be synchronised better with the laying equipment. The trenching gear (plough or jet) can be towed equidistantly behind the barge as she moves. Another solution would be the laying of cable during high tides when the barge can swim and the subsequent trenching at low tides when the barge lies on belly and can serve as hold point for the trenching equipment.

Another method developed for the installation of pipelines in tidal flats is the "bathtub" method [17]. A dredger prepares a wide flotation ditch for the laying vessel, which follows directly after the dredger. The laying vessel can float all the time in the ditch regardless of the tides. The laying vessel is directly followed by another floating barge, which refills the ditch. The leading dredger pumps its excavation soil by pipe to the tailing refill barge. The cable-laying vessel has always a pond of water to swim in.

#### 7.2.3.4 Trenching Depth

There is always a discussion about the ideal trenching depth providing sufficient protection at reasonable cost. While it is obvious that a deeper trenching provides a better cable protection, the relationship between trenching depth and protection level is not linear, though. In some cases, the protection level would not increase below a certain trenching level.

While a large burial depth can offer a better protection (up to a certain degree), it comes with some important disadvantages.

Slower process, need for heavier equipment, considerably higher costs, higher risk for equipment failure and cable damage, less efficient cable cooling, larger difficulties when the cable must be retrieved for repair, or decommissioning.

The optimum burial depth can vary considerably along the cable route depending on protection requirement and soil conditions. A mathematical model for the detailed planning of the degree of protection for cable route sections with different properties and risk patterns is suggested in [18]. The burial depth is discussed further in Chap. 8. It should be noted that the designed trenching depth not always can be achieved. Unexpected sub-bottom soil conditions can render the chosen trenching equipment insufficient.

The best protection for the cable is sub-bottom burial out of the reach of fishing gear. A burial depth of 1–1.5 m is a good protection against most of fishery threats. Trawling and fish dredging equipment do seldom reach depths exceeding 1 m. However, as sea bottoms can be volatile, a good sand cover might disappear after years or even months leaving the cable exposed and vulnerable for fishing gear aggression.

The choice of trenching depth might be influenced by the knowledge of future developments in the cable route such as planned harbour dredging, ship lane dredging, or future pipe installation.

#### 7.2.3.5 Other Protection Methods

There are also pre-laying trenching methods using dredging or excavating. The cable trench is prepared before the laying vessel brings the cable. The necessary furrow in the seafloor can be prepared by different machinery. Cutter-suction dredgers can work in a wide range of soils from mud to soft rock [17]. They can even cut an access ditch for themselves in shallow waters or mudflats. Here, also backhoes mounted on a small barge can operate.

If the seafloor is rocky or otherwise too hard for ploughing/jetting, the cable trench can be prepared before laying by cutting wheels, cutting chains, or other mechanical disintegrators. This is costly, slow, and only suitable for limited length of the cable route (Fig . 7.26).

In many cases, none of the methods mentioned here is feasible for one of the following reasons:

- Long rocky stretches which cannot be avoided
- Crossing of the cable over other cables or pipes
- The sand layer is too thin to achieve the specified burial depth
- Beach area does not allow for access with large vessels.

**Fig. 7.26** Wheel cutter for rocky seafloor. Cable trench cut into calcarite seafloor (Courtesy of LD Travocean, France)

For these cases, some other protection methods have been developed: an artificial cover protects the cable. The cover can consist of:

- Cast iron half pipes to be assembled onto the cable after laying. Mostly used in near-shore waters and applied by divers
- Concrete slabs
- Cement bags or mattresses
- Rock dumping.

Concrete slabs, cement bags, and mattresses can be applied from the sea level though for some projects purpose-built ROV have been developed to deploy mattresses over longer lengths.

Rock dumping is a technology well-known from protection of submarine pipelines of any kind and size. Purpose-built vessels carry a shipload of rocks over the laid cable and discharge the load. On some barges, the rocks are simply pushed over the side of the barge (side-dump). In a bottom-door vessel or split-barge the hold is opened underneath, and the load of rocks is falling down altogether. For the protection of cables, this method is simple and quick but wasteful. Better control and a more beautiful protection job can be achieved with discharge through a flexible fallpipe. Modern rock-dumping vessels are equipped with high-sophisticated control and monitoring instrumentation, e.g. DP systems.

A peculiar protection method was used for the 1967 SACOI HVDC cable between Italy and Corsica. After too many encounters with fishing gear, a number of heavy steel ropes were run next to the cable, resting on blocks at a certain height above the seafloor [14]. Their purpose was to prevent fishing gear from entangling with the power cable.

The strangest heard-of protection method, the alignment of scrapped busses along an Istanbul submarine power cable, could, however, not be confirmed.

### 7.2.4 After-Installation Protection

Even after the successful and completed installation of submarine power cables, the protection can be maintained and improved by active measures. Unfortunately, the after-installation protection measures are neglected in many submarine power cable projects.

Since the overwhelming majority of cable damages are caused by human activity, it is a good idea to keep people away from the cable. This is an information task. People must be told that there is a cable, and that it is dangerous or at least inconvenient to get in touch with it.

Beach warning signposts should be erected where applicable. Also, information on the cable position must be given to all issuers of sea charts, be it national marine agencies, fishing authorities, etc. National sea chart authorities would issue "Notices to Mariners" alerting on newly installed submarine cables. It is important to entry the submarine power cable into any sea chart and registry imaginable. The cable information should be disseminated to pipeline operators, harbour authorities, meteorological, and hydrographical agencies. Military authorities should be informed in any case even if their present operational zone does not include the cable route. London-based "International Cable Protection Committee" (www.iscpc.org) keeps cable registers mainly for the telecom cable industry but would also include submarine power cables into their charts.

The dialogue with fishermen and their organisations is crucial. Fishery authorities and fishermen's professional organisations are usually willing to help to spread information on submarine power cables. Every fisherman and mariner should know that it is a bad idea to try to recover an entangled gear or anchor from a cable by brute force. Mariners should know that it could be very dangerous to encounter a submarine power cable. The loss of at least three fishing vessels and several men of their crew has been attributed to hooking of submarine cables [19]. Every fisherman should also know that the cable owner would happily compensate him for a lost fishing gear rather than to repair a cable that has been damaged when the fisherman tried to recover that gear. It is also advisable to circulate information and education to fishing exhibitions and schools for fishermen and future mariners. In less developed areas, it can be profitable to provide local fishermen with easy-to-understand charts, free of charge, showing the position of the cable.

The Long Island Power Authority has hired local fishermen to guard their cable installation works. By this the local fishing community got aware of the cable existence.

Another way to protect the cable after laying is to monitor ship movements close to the route. The Vessel Monitoring System (VMS) specially designed for fishing vessels, or the AIS for all vessels, provide live data on ship identity and movements. By action of a marine authority or a specialized security company, the vessel can be warned upon approach to the cable corridor. Active ship movement monitoring can also be incorporated into harbour surveillance and other systems. Position and movement data may even be used in court proceedings when cable damages are dealt with. Many useful tips for the after-laying protection of submarine cables are given in [20].

Sea and air patrol of the cable route are efficient methods to keep possible hazards out. At least at times, the Cook Straight HVDC cable and other cable routes have been guarded regularly by sea patrols. Still, this service is expensive. Instead of establishing an own patrol service, the cable operator might enrol on other existing patrol services. Possible offenders can be identified and warned prior to cable damage. In case of damage, the offender can be identified for further claiming.

## 7.3 Appendix: The Catenary Line

When the CLV lays cable in calm waters with constant speed over a horizontal seafloor, the cable is following a catenary line from the exit point on the laying wheel down to the touchdown (TD) point on the seafloor. Actually some more simplifications are necessary to achieve a catenary line:

1. The cable has no bending stiffness
2. The cable does not experience any drag while moving through the water
3. The cable has a uniform weight per meter.

The bending stiffness of submarine power cables can be neglected in most cases except for very shallow waters. Given the slow laying speed the drag forces exerted onto the cable may also be neglected (in contrast to telecom cables, which are paid out at much higher speed).

In the coordinate system of Fig. 7.27 the catenary line can be expressed as:

$$y = a \cosh\left(\frac{x}{a}\right) \tag{7.7}$$

where $a$ is the catenary parameter

In Fig.7.28, the real laying situation is included in the diagram. Note that the seafloor is not at $y$=0 but at the vertical coordinate $y$=$a$. The angle $\phi$ is expressed by the derivative of Eq. 7.7:

$$\cot \varphi = \frac{\partial y}{\partial x} = \sinh\left(\frac{x}{a}\right) \tag{7.8}$$

The coordinates of the point where the cable leaves the laying wheel are $x$=$L$ and $y$=$H$+$a$. From Eq. 7.7 follows:

$$y = H + a = a \cosh\left(\frac{L}{a}\right) \tag{7.9}$$

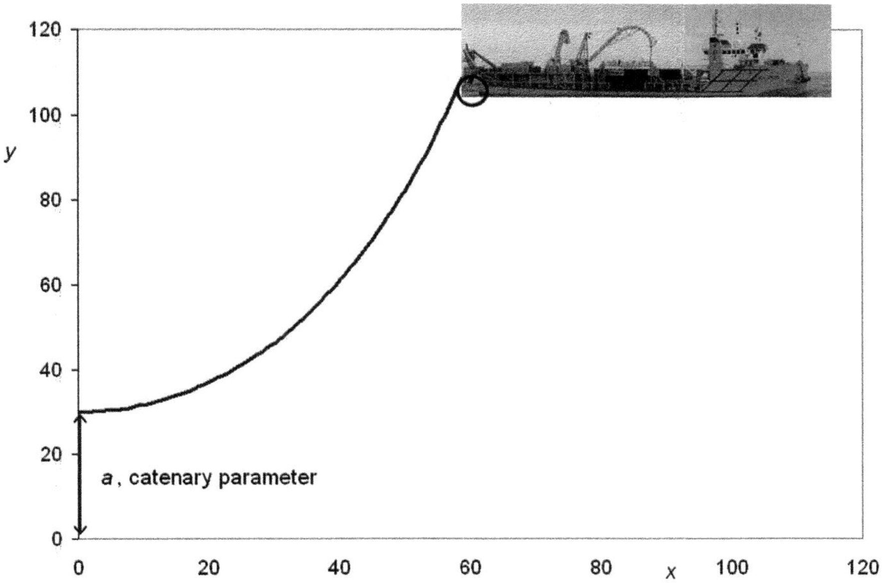


**Fig. 7.27** Catenary curve of the cable under the laying vessel. Here the catenary parameter $a$ has the value $a$=30

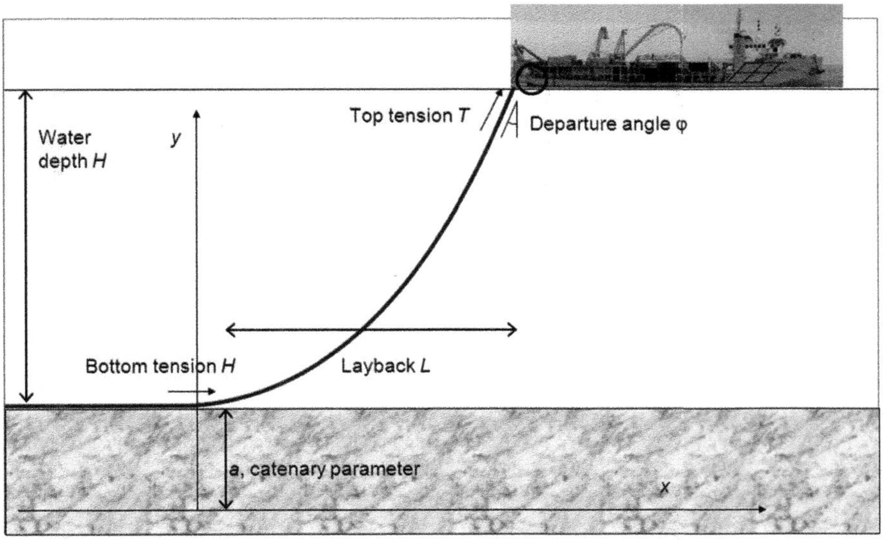


**Fig. 7.28** Catenary curve of the cable under the laying vessel

The inclination of the curve at the same place is equal to the departure angle $\phi$. During laying of the cable, the departure angle $\phi$ can be monitored and the water depth $H$ is known. In the following we calculate all relevant entities from the known values. From Eq. 7.7 follows:

$$\cot\varphi = \sinh\left(\frac{L}{a}\right) \text{ or } \frac{L}{a} = \operatorname{arcsinh}(\cot\varphi) \tag{7.10}$$

That means that $L/a$ can be calculated from the observed departure angle. Now, we put $L/a$ into Eq. 7.9 and solve for $a$:

$$a = \frac{H}{\cosh\left(\frac{L}{a}\right) - 1} \tag{7.11}$$

Now, the catenary parameter $a$ is calculated, and we can continue calculating the other important entities.

$$T_0 = w \cdot a$$

where $T_0$ is the bottom tension and $w$ the cable weight per metre in water. The layback $L$ can be calculated from Eq. 7.10. The total length $s$ of the suspended cable (from wheel to touchdown) is:

$$s = a \sinh\frac{L}{a} \tag{7.12}$$

The important top tension $T$ is:

$$T = \sqrt{T_0^2 + w^2 s^2} \tag{7.13}$$

Finally the bending radius at touchdown is:

$$R_o = a \cosh^2\frac{L}{a}. \tag{7.14}$$

## References

1. McBeth J (1974). The Use of Manned Submersibles for the Burial of Submarine Telecommunications Cable, OCEANS, Vol. 6, pp. 209–214, August 1974.
2. Bartnikas R (Ed.) Power and Communication Cables, IEEE Press Series on Power Engineering, ISBN 0-7803-1196-5.
3. Det norske Veritas DnV-RP-F109: On Bottom Stability of Offshore Pipeline Systems.
4. Schmager G (2003). Maritim-meteorologische Informationen für Seeverkehr und Offshoreindustrie. In Neues Vom Wellenatlas, F-M Chmielewski et al. (Ed). Beiträge zur Klima- und Meeresforschung, Berlin und Bayreuth, pp. 169–177.
5. Ainsworth T (2006). When do Ocean Waves become 'significant'?, In Mariner Weather Log, Vol. 50 No. 1, April 2006, published by the National Oceanic and Atmospheric Administration (NOAA), USA. (http://www.vos.noaa.gov/MWL/apr_06/waves.shtml).

6. Soares C G et al. (2001). Probability Distributions of Wave Heights and Periods in Measured Two-Peaked Spectra from The Portuguese Coast. Proceedings of OMAE.01, 20th International Conference on Offshore Mechanics and Artic Engineering, Rio de Janeiro, Brazil, June 3–8, 2001.
7. http://www.marine.tmd.go.th/Part-II-Chap1.pdf, published by the Thai Marine Meteorological Centre.
8. Kjeldsen S P (1997). Examples of Heavy Weather Damages Caused by Giant Waves, Bulletin of the Society of Naval Architects of Japan, Vol. 820-1997/10, pp. 24–28.
9. http://www.containerhandbuch.de/chb_e/stra/stra_02_03_03.html, in English.
10. Dobie T G (2003). Critical Significance of Human Factors in Ship Design. Proceedings of the 2003 RVOC Meeting, 8–10 October, 2003, Large Lakes Observatory, University of Minnesota, USA.
11. Yamamoto T et al. (1964). Matsushima Oil-filled Submarine Cable Line Construction, Sumitomo Electric Technical Review, No. 4, July 1964.
12. Berthelsen G (2008). High Reliability – Low Cost Submarine cable Deployment. Submarine Telecom Forum published by Unity Business Solutions on the Internet, September 2008. http://www.subtelforum.com/articles/wp-content/wfn_40.pdf
13. Allan P G (2001). Hydrographic Information and the Submarine Cable Industry. Proceedings Hydro 2001, Norwich, UK, March 2001.
14. Cigré Working Group 21.06 (1986). Methods to Prevent Mechanical Damage to Submarine Cables, Working Group 21.06, Cigré Session 1986.
15. Allan P G et al. (2001). The Selection of Appropriate Burial Tools and Burial Depths, Displayed at SubOptic, Kyoto, May 2001.
16. Hettinger F et al. (2006). Cable and Pipeline Burial at 3000 Meters, Oceans 2005, September 19–23, Washington, DC, USA.
17. Palmer A C and King R A (2006). Subsea Pipeline Engineering, PennWell, Tulsa, Ok, USA.
18. Nakamura M et al. (1992). Reliability Analysis of Submarine Power Cables and Determination of External Mechanical Protections, IEEE Transactions on Power Delivery, Vol. 7, No 2, pp. 895–902, April 1992.
19. Report of the Inspector's Inquiry into the Loss of the Fishing Vessel WESTHAVEN AH 190 with Four Lives on 10 March 1997 in the North Sea, November 1998, Marine Accident Report 4/98, Marine Accident Investigation Branch of the Department of the Environment, Transport and the Regions: London.
20. ICPC Recommendation No. 6 (2007). Recommended Actions for Effective Cable Protection (Post Installation).

# Chapter 8
# Damages and Repair

## Contents



## 8.1 Damages

Submarine cables have been stricken with damage and failures since the early days. Although one may think that submarine cables, in their depth, are hardly accessible for human activities, experience from 150 years of submarine cable history has shown that most cable breaks are caused by humans. It started when a French

T. Worzyk, *Submarine Power Cables*, Power Systems,
DOI 10.1007/978-3-642-01270-9_8, © Springer-Verlag Berlin Heidelberg 2009

fisherman caught the first Dover-Calais submarine telegraph cable in 1850, only days after its installation. According to some accounts, he believed it was a delicious new sort of seaweed; according to others, he took the conductor for gold.

The reader of this chapter should not conclude that the installation and operation of submarine power cables is an unpredictable gambling with unavoidable incidents beyond control. In contrary, a thorough analysis of the installation and operation conditions gives the preconditions to

(a) take measures to reduce the operational risk substantially
(b) arrive at a correct perception of the character of the remaining risks
(c) enjoy a high availability of the submarine cable system.

During the history of submarine cables, strategies were developed to protect submarine cables from damages. Today, submarine power cables belong to the most reliable components of electric power systems. Still, utilities need to be prepared for the unlikely case of a submarine power cable failure. Unfortunately they cannot just go there and replace the faulty piece of cable. Or can they?

### *8.1.1 Causes of Damages*

Fishing gear and anchor damages are accused for most of submarine cable failures, both for telecom cables and power cables. Figure 8.1 shows the cause of damages to telecom cables in the Atlantic [1].

The distribution of damage causes can change considerably in different waters, and for different cable types. Cables in shallow coastal waters have a higher risk of damages by fishing gear than cables in harbour areas. Power cables will not be affected by fish bite as much as slim telecom cables. However, Fig. 8.1 can illustrate the variety of damage causes. Beside the causes given in the figure, others have

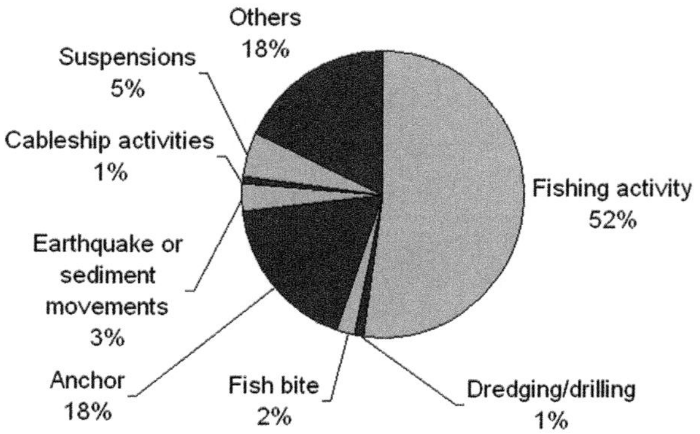


**Fig. 8.1** Cause of damages to telecom cables in the Atlantic

been reported, including cable damage by dynamite fishing off the Philippines and elsewhere. In the following, some hazards to submarine power cables are discussed.

### 8.1.2 Statistic Distribution of Damages

The statistical assessment of submarine cable damages is of great interest for cable designers as well as forcable operators and insurance companies. Cable link availability and the risk of repair costs are critically depending on the expected frequency of cable damages along the specific submarine cable route. Knowing the largest hazard for the submarine cable, the design engineer can optimise the design of the cable armouring and an appropriate protection of the cable. Using the expertise of the cable manufacturer and the installation company, the cable operator can balance the costs for different levels of armouring and protection against the economic risk of cable damage. The repair of submarine cable fault is costly, and often takes considerable time because repair vessels are not always available, and weather conditions may impede repair attempts for weeks or months. The possible loss of income is another reason for a better cable protection.

Comprehensive studies on statistics of submarine cable damages are available from the telecom cable industry as they install vast amounts of submarine telecom cables. Only between 1997 and 2001, 350,000 km of submarine telecom cables have been installed [2]. The annual failure rate of submarine telecom cable networks can vary considerably between years [3].

It is not quite clear if this variation also applies to submarine power cables that tend to have a larger diameter and greater strength than telecom cables, and hence are less vulnerable to minor aggression. But as fishing industries employ larger units, and ship anchors grow larger, even the biggest submarine power cable may be at risk.

The Cigré study on submarine power cable damages published in 1986 [4] does unfortunately not specify the type of external violence that caused the investigated cable failures. The most important factor, altering a bad failure statistics into a good one, is the improvement of the cable protection.

### 8.1.3 Damage by Fishing Equipment

To understand the risks of damages to submarine cables some words should be mentioned about the most common fishing techniques, which may affect submarine cables.

The speed of Otter trawling (Fig. 8.2) is about two up to four knots. The trawl doors (= otter boards) are large flat pieces of steel, designed to keep the net open horizontally, while it is towed over the seafloor by the trawler. The trawl doors slide over and below the seabed in order to keep the net down, and thus cut – in soft grounds up to 30–40 cm deep – into the seabed. The weight of the trawl doors can be

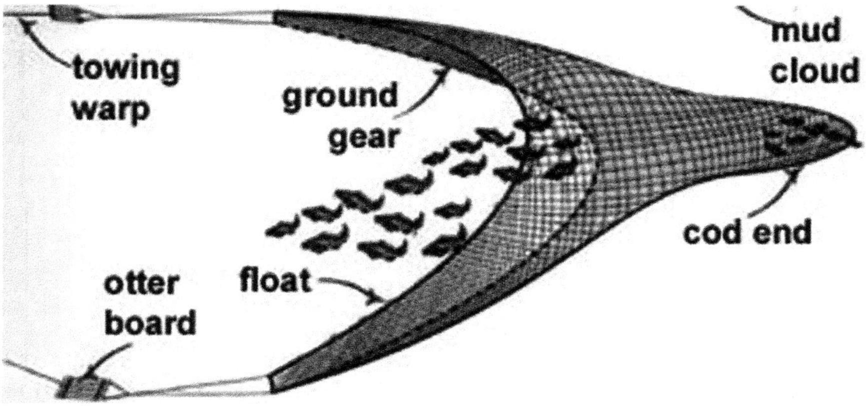


**Fig. 8.2** Otter trawl net configuration
Source: Wikipedia/public domain

up to 3.5 MT. Some trawlers tow two or three nets at the same time. Under unusual conditions (break of the connection bridle between trawl door and net, or badly rigged set-up) the door can cut down into the seafloor much deeper than normal.

Beam trawling involves a horizontal beam to keep the net open, rather than trawl doors. The beam heads also can entangle with cables. Beam trawls caused over 100 telecom able faults in the North Sea between the mid-1960s to mid 1970s [5].

The force impact on a cable when hit horizontally by a 1900 kg trawl door at a trawler speed of 2.9 kn, has been estimated to 11 t [5]. Other components of the trawling gear, such as bridles, shackles, and rockhopper discs also have the potential to fault the cable. However, the largest hazard for the cable comes up when the trawler crew tries to recover the entangled fishing gear. Modern large trawler can develop high winch and bollard pull forces which can destroy almost every hooked submarine power cable.

Dredge fishing is designed to stir up the bottom sediment in order to harvest shellfish. The search for clams, scallops, and crabs may involve dredging equipment going 450 mm deep into the seabed. Even here, unusual conditions (such as rocks ahead) can push the dredging equipment deeper into the sediment.

Under normal conditions neither trawling nor dredge fishing is expected to disturb the seabed more than 0.5 m in depth. For a submarine cable, that is buried deeper than 1 m into the seabed, the risk of interference is low. However, the bathymetry is not always stable. The seabed can move with time and expose the cable for external threats.

Another fishing method, bottom set fixed fishing, uses stationary nets to catch fish in the move, driven by either fish migration or fish in streaming waters, e.g. in tidal currents. The nets are fixed to the sea bottom by anchors or weights. With stow net fishing, a method often used in Korean and Japanese waters, anchors up to 1.5 t are used to keep both the net and the attending vessel at station. These anchors can penetrate a sandy sea bottom down to –1.5 m, or a soft sea bottom down to

–2.7 m [3]. The nets are positioned in places where the tidal currents are strong in order to catch large fish swarms. If the anchors hook into the submarine cable, the strong tidal current can subject the cable to considerable forces. Tired of many cable damages from fishing activities in the Japanese Seto Inland Sea, the local power utility Chugoku Electric Power Co. installed a new heavily armed 66 kV submarine cable in 1992. When it was hooked later, the outer armouring suffered some damage, but, after testing, the cable was considered fit for further operation [6].

Fishing is a billion-Dollar industry and makes a living for many millions of fishermen all around the world, who would just not easily accept being told: Please don't fish close to my cable! Methods to protect submarine power cables from fishing gear damages are described in Chap. 7.

### *8.1.4 Damage by Anchors*

According to Fig. 8.1, anchor damages account for the second largest single cause of submarine cable damages. Given the larger diameter and strength of submarine power cables in contrast to the telecom cables that Fig. 8.1 was compiled for, the anchor damages most probably are at level with fishing gear damages or even exceed them.

At first sight, it seems incomprehensible that a commercial vessel is able to drop anchor onto a cable that is depicted in sea charts. Any commercial vessel should be equipped with the latest charts and avoid cable areas when mooring. However, reality is different.

A large category of anchor damages is caused by mishaps. Anchors are being dropped involuntarily. Reasons can be badly secured anchors, which go down unintended. Master and/or crew affected by alcohol or other drugs might be responsible for anchor dropping. Reportedly, vessels have sailed along with their anchors dragging behind.

Another category of anchor aggression stems from emergency anchoring. Ships, large or small, can loose control of manoeuvrability by electric power black-out, engine malfunction, drop of rudder function, leakage, or for other reasons. Then the ship Master or Commander-on-Duty can drop anchor to save life or ship. Saving life has the highest priority under any circumstances. On other occasions, anchors have been dropped into cables because the harbour tugs were on strike and the vessel had to stop somehow. Ironically, anchors from cable laying barges have cut into cables being laid by the same barge only a little earlier. Drop of anchor has happened, and will happen, everywhere where ships manoeuvre: in harbour basins, in ship lanes, or in open seas (Fig. 8.3).

A number of different anchor types are being used with weights up to 30 t for the largest ships. The anchors of "Queen Mary 2" weigh 23 t. Different anchor shapes have different efficiency and behave different in their contact to the sea bottom.

The risk of anchor damages of a given cable in a given cable route can only be assessed with statistical methods. The assessment starts with the analysis of ship traffic in the investigated cable area. National shipping authorities often can provide

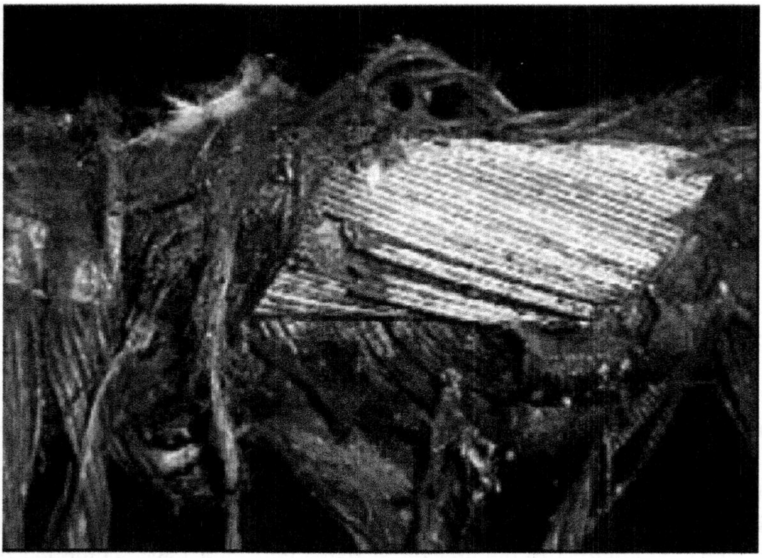

**Fig. 8.3** Anchor damages on the 120 mm Baltic Cable (Courftesy of Baltic Cable AB, Sweden) [7]

statistics on the number of vessel crossings over the investigated cable route, and the size distribution of the ships. The next step would be to determine the anchor size distribution. From the compilation of a large number of ships and their anchor weight, Luger [8] proposed the following empirical relation between anchor size and ship dead-weight:

$$y = 7e - 3 \cdot x^3 - 6e - 7 \cdot x^2 + 0.1635 \cdot x + 2162.2 \qquad (8.1)$$

where $x$ is the ship's dead-weight in metric tons and $y$ is the anchor mass in kg.

Using this relation, one can establish an anchor size distribution from the ship size distribution in the cable route. Figure 8.4 shows the expected accumulated anchor size distribution, calculated for the Kadetrenden strait between Denmark and Germany [7]:

It appears that about 98% of anchors are at 8000 kg or below, and more than 60% are below 4000 kg. The ability to damage a buried submarine cable increases with anchor weight by three reasons: A large mass has a larger damage capability at direct hits. A heavier anchor can penetrate the sea bottom deeper than a lightweight anchor. Finally, a heavy anchor belongs to a heavier vessel which can drag and damage with greater force than a smaller ship with a smaller anchor. Not all anchors are equally dangerous. In sand or clay, anchors usually can dig down only as much as a fluke length; while in soft mud, they can go down 3–5 fluke lengths (Table 8.2).

Another source [9] quotes relationships between ship size, anchor weight, and penetration depth for Japanese waters. Unfortunately, the authors do not specify which ship tonnage unit is meant. The anchor weights from Table 8.1 and Eq. 8.1

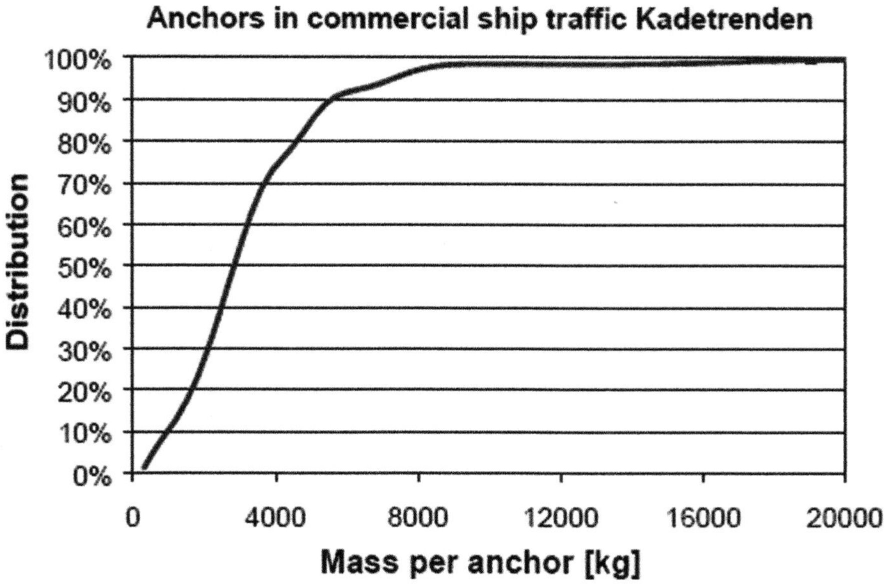


**Fig. 8.4** Accumulated frequency of anchor size in the Kadetrenden strait

**Table 8.1** Relationship between ship size, anchor weight and penetration depth according to [9]

| Ship weight (tons) | Anchor weight (tons) | Penetration depth in mud | Penetration depth in sand |
|---|---|---|---|
| 1000 | 1.0 | ~ 1.0 m | ~ 0.5 m |
| 5000 | 2.8 | ~ 2.0 m | ~ 1.0 m |
| 15,000 | 4.6 | ~ 3.0 m | ~ 1.5 m |
| 50,000 | 8.2 | ~ 4.0 m | ~ 2.0 m |
| 100,000 | 12.4 | ~ 5.0 m | ~ 2.5 m |

**Table 8.2** Relationship between ship size and penetration depth in firm ground (sand/clay) according to [3]

| | Anchor type | |
|---|---|---|
| Ship weight (Gross tonnes) | Danforth/Moorfast penetration depth | Stockless/AC 14 penetration depth |
| 2000 | ~ 1.5 m | ~ 1.2 m |
| 4000 | ~ 1.7 m | ~ 1.35 m |
| 70,000 | ~ 1.9 m | ~ 1.45 m |
| 100,000 | ~ 2.1 m | ~ 1.5 m |
| 130,000 | ~ 2.2 m | ~ 1.5 m |

correlate quite well, but underestimate the real anchor weight of "Queen Elizabeth II" by a factor of three.

Using some further probability figures, Christensen summarizes his results in Fig. 8.5:

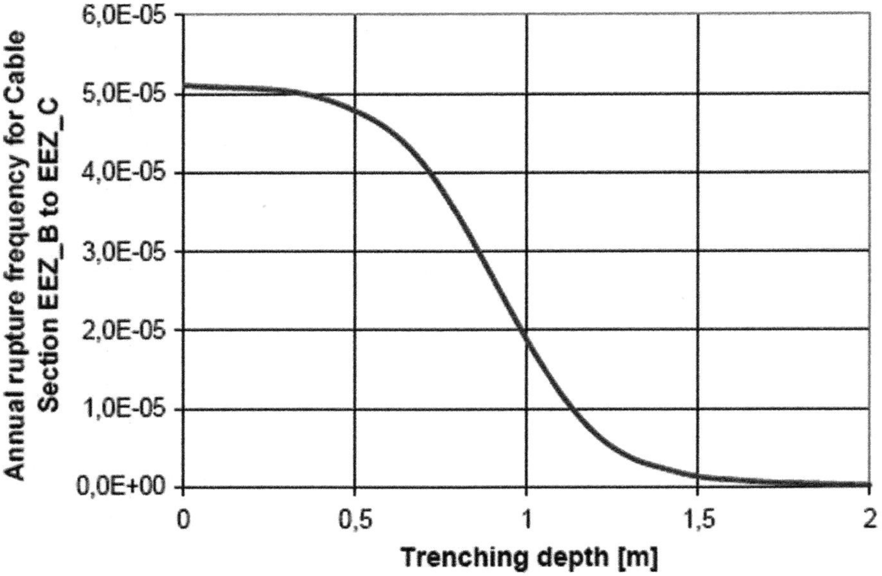


**Fig. 8.5** Annual rupture frequency for a submarine cable in the Kadetrenden [7]

It becomes clear that a burial depth of 1.5–2 m is a very good measure to avoid anchor damages in these particular waters. A similar curve can be expected for most shipping areas. Similar risk assessments for other cable routes should be performed, based on the local traffic pattern and density.

### *8.1.5 Damage During the Installation*

While many failure statistics account for failures during operation, they usually do not include damage to the cable that happens before commissioning. Cable damages during the installation might call for expensive and time-consuming repair operations.

Experience has shown that almost anything imaginable and unimaginable can happen during installation. The following lists some incidents known from various submarine power cable installations. The list aims by no means at scaring cable owners but at helping them to prevent some sources of trouble.

*Loss of DP*. The CLV is may be unable to keep position due to storm, waves or currents. The uncontrolled vessel movements can introduce inappropriate strain, bending, or choking of the cable. A case where the laying barge moved backwards unintendedly introducing backwards overbending into the already laid cable, belongs to this category.

*Anchoring damages*. It so happened that the anchors used for the mooring of cable laying barges damaged cables that have been laid in previous runs.

*Kinks*. Insufficient coordination of CVL forward motion and cable pay-out may cause kinking of the cable at the seafloor or over the seafloor. In the best case, the cable can remain as it is laid and only the subsequent protection by plough or jetting is rendered impossible.

*Loading/re-loading.* Cables can be damaged during loading onto the cable vessel, or during transfer between a transport vessel and the cable laying vessel.

*Protection.* Inappropriate trenching equipment or methods are able to damage the cable by squeezing, deformation or overbending.

*Emergency cut.* Extreme situations may emerge during laying/jointing that call for immediate emergency cutting of the cable on-board the CLV in order to avert situations dangerous for crew or vessel. An emergency cable cut usually necessitates time-consuming repair operations.

### *8.1.6 Other Damage*

Anchors and fishing activities account for the largest portion of submarine cable damages. A minor share is caused by other factors, this portion being smaller for submarine power cables since they are stronger than telecom cables.

Many damages are induced in cables installed with free spans. Sometimes difficult submarine areas with rocky areas and large outcrops cannot be avoided. Even with the most suitable installation and protection methods being used, the cable may end up hanging in free spans over submarine obstacles. This situation combined with water currents can easily lead to oscillations induced by vortices, a phenomenon known as vortex-induced vibrations (VIV), or, somewhat misleading, "vortex shedding". The oscillations can cause strong chafing at the contact points of the cable to fixed ground, and premature fatigue of the lead sheath. The water currents that may cause harmful oscillations might stem from [10]:

- atmospheric pressure gradients, storms
- tidal currents
- circulation currents (e.g. Gulf Stream and spin-offs of the large marine current systems)
- longshore currents – run in parallel to the coast line and are driven by diffracted waves.

Currents up to 7 kn in cable routes have been measured, e.g. in the Öresund strait between Denmark and Sweden [11]. The first cable system laid between the North and South New Zealand islands was damaged several times in free spans where the cable was suspended between outcrops. A more detailed description of vortex-induced vibrations is given in Chap. 3. The advice "do not install cables in free spans" is given in many accounts on submarine cable projects.

Not only anchors, but also direct hits by ships have damaged submarine cables. The ferryboat "Nils Holgersson" ran aground outside Travemünde / Germany due to a loss of power on March 16th, 2002. The 450 kV HVDC "Baltic Cable" was damaged by the keel and the power link was interrupted. When the freighter "Roy A. Jodrey" sank with 20,000 t of iron ore under the eyes of the Coast Guard in the St. Lawrence River, she cut off the submarine power cable supplying the Coast Guard station, leaving them in the dark [12].

Oceans and peripheral seas are subject to geo-hazards implying potentially damaging phenomena such as landslides, volcanic activities, gas hydrate destabilisation, etc. Movements of the sea bottom due to submarine landslides or earthquakes can destroy submarine cables by direct impact or by land slide induced turbidity currents (Suva 1953 [13], Taiwan 2006). Volcano activities destroyed submarine power cables at the Heimaey eruption (Island, 1973) and elsewhere. Landslides and other movements of the seafloor can even be caused by annual formation/deformation of gashydrates, a material that can exist only in certain waters under certain pressure and temperature conditions below approx 200 m [14]. Gashydrates have the potential to change structure and mechanical strength during the seasons and over the years. Their change may impose a hazard to subsea installations.

Shark bite has caused frequent outages on early submarine telecom cables in the Canary Island region but is not believed to be a serious problem to armored submarine power cables today.[1]

A charting of the estuary of the river of Thames shows a large number of shipwrecks, many of them from famous sea battles of the past centuries. Some of the wrecks are painted red in the chart, indicating that they are carrying unexploded ammunition. This is not a good place for submarine cables.

### *8.1.7 Spontaneous Damage*

Electric insulation systems have a limited lifetime. A number of ageing phenomena affect the insulation where temperature and electric stress are the dominant factors. Different ageing mechanisms are discussed in references [11, 13] of Chap. 2. Spontaneous damage, or internal failure, occurs extremely rarely in submarine power cables. The typical lifetime for cable insulation ranges from a few years (cables manufactured or installed with inappropriate methods) up to well over 50 years for some paper-insulated cables.

The proper function of the insulation is sometimes obstructed by other causes than ageing. Fatigue cracks in the lead sheath can result in loss of oil or ingression of water. A pressure drop in the oil-feeding station ashore easily detects larger oil losses in oil-filled cables. Small leaks can exist over weeks before the oil loss is noticed. Water intruding into the insulation deteriorates the insulation properties and will eventually cause a breakdown. The hydrophobic properties of impregnating oils/compounds in paper-insulated cables delay the propagation of the water

[1]The lethal encounter of a Great White with a submarine power cable in the movie *Jaws II* was not voluntary, nor realistic.

considerably. Laboratory tests with mass-impregnated d.c. cables have demonstrated that it can take weeks or even months from the first ingress of water until the breakdown finally occurs. However, the insulation of low and medium-voltage submarine cables can be made tolerant against intruding water because of the low electric field and purpose-made insulation resins.

Another failure mechanism observed in older mass-impregnated cables is caused by the depletion of the impregnation compound in elevated parts of the cable, e.g. on the top of a ridge. Under unfavourable conditions, the compound is suspected to migrate "downhill" leaving unfilled insulation behind. However, to the knowledge of the author, no such case has been reported within the last 10 years. Mass-impregnated cables are not suitable for vertical installation, and manufacturers have different opinions on the maximum allowed inclination angle for their mass-impregnated cables.

In most HVDC cables, the lead sheath is electrically connected to the armouring in regular distances, the connections being made through the plastic sheath over the lead sheath. At least in some cases, faulty designed ground connections are suspected to have caused internal sparks in the sheath area eroding the lead sheath and leading to breakdown.

The new Cigré investigation (2009) on service experience with submarine cable systems embraces 7000 reported circuit kilometres of submarine power cable [15]. Of 49 faults reported in total for the time span 1990–2005, only four faults were identified as "internal", i.e. spontaneous faults. Three of these faults struck the same cable installation.

### 8.1.8 Failures of Joints

According to the 1986 Cigré study, a noticeable part of the damages were located in submarine or beach joints. Unfortunately the study does not specify the exact failure causes. Three major fault causes for submarine cable joints are known:

- inadequate design
- poor joint assembly work on-board
- adverse weather condition during assembly.

Many of the cables evaluated in the study were oil-filled cables. The joints of these cables are more complicated and demanding than joints of solid-type cables (XLPE or mass-impregnated). The first Cook Strait cable suffered a number of failures in transition joints between different conductor sizes, so that the new Cook Strait cable was specified to have only one conductor size. In the past decade submarine cable accessories have become much more reliable because of better understanding of mechanical stresses, better design, and more sophisticated installation practice. The 2009 Cigré study [15] reports only four joint failures out of 49 failures in total in 7000 km submarine cable.

## 8.2 Repair

The repair of submarine cables is one of the most demanding tasks of submarine cable engineering. Not only must the fault be found and identified in water depths often inaccessible to divers, also the damaged cable must be made accessible for repair. Attempts were made to repair damaged cables on the seafloor. Leaking lead sheaths in the 138 kV Long Island cable system were repaired using external sealing cassettes, which were attached to the cable by divers [16]. Another submarine repair attempt was made using submersible workshops to repair oil-leakage from submarine power cables [17]. Also, for the IFA 2000 HVDC link between France and UK, repair methods based on manned submarine vehicles were devised [18]. Neither method had been pursued further.[2] In general, the damaged cable must be brought to the surface safely. Once it is onboard of the repair vessel, the cable must be cut back to undamaged fresh cable and jointed on a possibly rolling and pitching repair deck when the heavy cable hanging over board may expose the equipment to too strong a tension. Adverse conditions have forced some repair operations to terminate in emergency.

Submarine cable repair requires good preparation. "Red alert" plans should be available for each cable link including repair method descriptions, service/maintenance contracts, vessel and crew availability plans, etc.

### *8.2.1 Spare Cable*

Almost every repair job requires that the cable would be lifted on board the repair vessel. In most cases, the cable lays on the sea floor on a straight line and must be cut before lifting the ends. For the repair, a spare cable must be jointed between the two recovered cable ends. The length of the spare cable must at least cover the double water depth, an additional length to establish the catenary lines for the cable, plus the amount needed on board of the vessel in the cable gantries, plus the length cut away during jointing, plus safety margins. Before jointing, the ends of the damaged cable would be checked for further damages and/or water ingression, which might require additional cut-back of the cable. If the jointing has to be re-done for any reason, further length of spare cable is required. The experienced repair operation manager will bring a long spare cable to the repair site rather than finding himself short of spare cable before the final joint is finished.

Few submarine power cables are standard products that can be procured within a week's time. Producing new spare cable takes months, at unit costs much higher than the original production. Therefore, the most cable operators/owners have a stock of spare cable, which has been produced together with the original submarine cable.

[2]The following question has been passed on in this context: Is it easier to train a diver to be a jointer, or to train a jointer to be a diver?

Should the cable operator wish to increase his stock of spare cable he sometimes can place an order at the supplier to be produced as "hang-on" to another cable manufacturing order. Spare cables are usually stored in quay warehouses for easy loading to repair vessels. Spare cables for submarine power cables can even be stored on the seafloor at suitable locations.

### *8.2.2 Repair Vessel*

The repair vessel may differ very much from the laying vessel, as there are different needs for load capacity and deck arrangements. A load capacity of a few hundred tons can be enough to carry the spare cable and all handling equipment. Also, the requirements for a repair joint and an installation joint are different even if the joints per se are equal.

The repair vessel has preferentially a large open deck with sufficient space for the jointing house, cable engines, winches, cranes, etc. It is quite common to mobilise a supply vessel or barge with the appropriate equipment for the job. A turntable or cable hold for the spare cable must be installed on the vessel as well as chutes to deploy the cable and joint over board. For more complicated repair operations, an ROV might be necessary. Other specialised vessels before the repair operation usually perform the de-burial of damaged cables.

For the insertion of a piece of spare cable into a broken cable, two different types of deployments are needed. The first joint would be made between one recovered cable end and the spare cable on-board. The joint and spare cable can then be paid out over the chute. The final joint between the other recovered cable end and the spare cable creates a loop of cable on-board that must be deployed in a different arrangement. A possible repair vessel lay-out is shown in the example at the end of this chapter.

As it is the case for the CLV, the repair vessel needs suitable DP and other navigational equipment depending on the character of the task. Repairing a 600 MW HVDC cable 100 km out in the raving sea requires a different set-up than the repair of an 11 kV cable in a calm bay of a subtropical archipelago.

### *8.2.3 Repair Crew*

Apart from the nautical crew of the repair vessel, also a specialist team of cable jointers is necessary. While there are many companies providing cable repair for land cables, there is only a limited number of teams skilled in the art of submarine cable repair. Most reputable submarine cable manufacturers have their own trained jointer teams. Given the large day rates of repair vessels, it is advisable to hire a sufficient number of jointer teams to get the work done on a 3-shift basis.

Many vessel operators require that jointer crews produce medical certificates and/or certificates on relevant safety and survival training before embarking.

### 8.2.4 Repair Operation

There is no standard cable repair operation for submarine power cables. Sad exemptions from this rule are those few cables that are affected by similar faults over and over again. However, there is a generic sequence of activities that can serve as a basis to design a specific repair task.

The generic repair sequence is the following:

- cut the cable at the fault site
- take up one cable end and joint it to the spare cable carried on-board
- lay spare cable and proceed to the location of the other cable end
- bring up the other cable end and joint it to the spare cable
- put everything back down to the seafloor.

A possible repair operation is described in a section at the end of this chapter.

Upon occurrence of a cable fault, a number of activities would start. As losses of income may be high for each day, it is an all-out effort:

- Fault location. The position of the cable fault can have strong influence on the required repair spread. The coarse location can be performed from shore sites with sufficient accuracy to plan the repair job. Faults in the splash zone, in shallow waters or in deep waters would require different repair equipment. Fine location requires the use of a search ship.
- Securing of repair vessel contract.
- Planning of repair operation. Repair methods are depending on water depth, weather and current conditions, cable dimension, joint design, and much more. The planning will result in the identification of necessary equipment and vessel modification.
- Mobilisation of repair vessel and equipment. Often the vessel needs to be modified with extra cable handling equipment, chutes, roller guides, etc.
- De-burial of the faulty cable portion
- Loading of spare cable and jointing kit
- Jointer crew embarks.

After all this, the repair vessel is ready to go. It is presumed that the cable jointing per se has been engineered and trained before the fault occurred. Experienced installation engineers will have prepared detailed repair plans including HAZOP (Hazard and Operability) studies. Daily meetings during the repair operation summarize achieved progress and coordinate next activities.

The duration of the repair operation depends critically on the availability of a repair vessel with suitable equipment, and the time needed to bring spare cable to the site. Small cables in calm waters can be repaired within a week or two, available spare cable provided. Large cables require heavier equipment and access to specialized vessels. If spare cable is available the downtime of the system can be between a few weeks and four months. Procurement of new spare cable can delay the repair operation by many months. In the 2009 Cigré study on submarine cable experience 19 of 49 reported faults could be repaired within one month [15].

Also, the weather conditions have a strong influence on the cable repair. Adverse weather conditions prohibit repair operations even with the largest repair vessels. Storm seasons can make it difficult to find suitable weather windows that hold up long enough.

## 8.3 Fault Location

A number of methods for fault location are available, each having its own possibilities and limitations. Depending on the characteristics of the fault and the design and configuration of the cable, the methods will have different prospects to successfully localize the fault. In case of a submarine cable fault, the cable operator often mobilises different methods simultaneously in order not to lose time, should some of the methods fail to identify the fault location.

Although there is a wide range of commercial fault-location equipment with ever more simplified user guidance available, the need for experienced engineers for the task must not be underestimated. The cable operator can make a prudent investment by sending some of his engineers to fault-location training courses.

Most often, a cable failure is being noticed by a voltage breakdown. This indicates a breakdown of the cable insulation and renders the cable inoperable. The damages range from a high-ohmic fault to a low-ohmic insulation damage or possibly a complete rupture of the cable.

In two- or three-core cables, the first step would be identifying the faulty cable core by a simple insulation test.

### *8.3.1 TDR*

The acronym TDR stands for "Time domain reflectometry" and is based on an electric impulse, which is sent into the faulty cable conductor. The easiest fault to detect is a complete cable break. The impulse travels down the conductor until it reaches the cable break. Here, most of the impulse energy is drained into the water, but a fraction is reflected with opposite polarity. The reflected impulse is travelling back to the shore cable termination, where the instrument records the impulse and shows it on a graphic display. Knowing the impulse propagation velocity, one can calculate the distance to the fault by measuring the travel time. Figure 8.6 shows a typical trace from a TDR measurement. The impulse propagation velocity $v$ can be calculated as follows:

$$v = \frac{1}{\sqrt{\varepsilon_0 \varepsilon_r \cdot \mu_0 \mu_r}} \tag{8.2}$$

with the symbols according to standard conventions. For mass-impregnated cables with $\mu_r = 1$ and $\varepsilon_r = 4$, the resulting propagation velocity $v$ is 150 m/μs, that

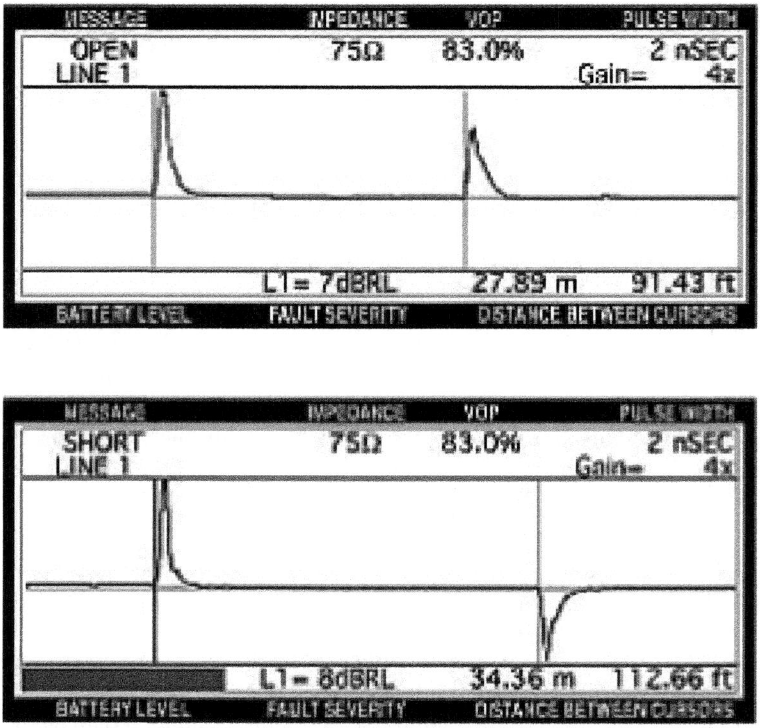


**Fig. 8.6** TDR reflections. Upper diagram: a reflection with the same polarity indicates a fault with OPEN (high impedance) tendencies. The reflection shown at the 2nd cursor is a COMPLETE OPEN. Lower diagram: a reflection with the opposite polarity indicates a fault with SHORT (low impedance) tendencies. The reflection shown at the 2nd cursor is a DEAD SHORT [19]

is half the speed of light. This value is independent of the dimensions of the actual cable. For one recently produced delivery length of mass-impregnated cable, a value of 141.26 m/μs has been measured, reflecting a $\varepsilon_r = 3.9$ rather than 4.0. Confusingly enough, in some publications and pamphlets, the value v/2 is specified as wave velocity because it is more convenient for the fault-locating engineer to multiply $v/2$ directly with the total travel time (forth and back) to find the fault distance.

As the real propagation speed of an impulse in a cable can be different from the theoretical speed according to Eq. 8.2, it can be prudent to determine the true velocity in the actual cable. This can be done directly after laying, by sending an impulse into the cable conductor and recording the reflection from the far-end cable termination if the far end is in the reach of TDR measurements. When the exact cable length is known, the propagation velocity can be calculated from the travel time. Also, $v$ can be determined by using TDR on an identical, healthy cable core with known length. Test engineers have reported that the propagation speed is different before and after installation. This has been attributed to different temperatures but the constants in Eq. 8.2 normally are not or very little temperature dependent. The difference in propagation speed has also been explained by the fact that the cable is reeled before installation and straight after installation.

Usually, rectangular impulses are sent out into the cable. The pulse length and the steepness of the pulse front are important for the spatial resolution, i.e. the accuracy of the location. An impulse with 10 ns pulse length at a propagation velocity of $v = 150$ m/μs can resolve distances down to $\Delta x = 1.5$ m. The impulse shape and impulse length is changing during the travel through the cable. This phenomenon is known as dispersion and is caused by wavelength-depending propagation speed of impulses. As a result, the rectangular impulse shape with a steep front, which has been sent into the cable, turns into a broader impulse with soft edges. The spatial resolution ability of this dispersed impulse is reduced. For this reason, distant cable faults can be located with lower accuracy than near faults. A longer impulse length from the TDR instrument brings more energy into the cable with better chance to record long-distance faults, at the expense of spatial resolution.

TDR actually measures impedance changes along the cable. These can also be generated by joints and defects.

TDR equipment is available in a large variety. Hand-held or desktop instruments with an impulse amplitude of a few volts are available. Other systems include impulse generators with some kilovolts of impulse voltage.

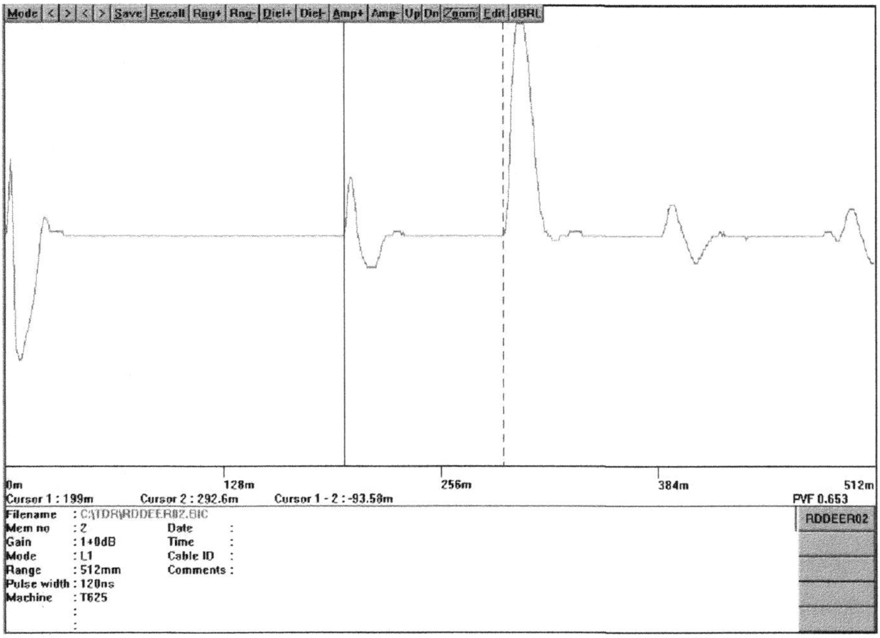


**Fig. 8.7** Analysis of a land cable TDR waveform [19]: Perfect trace, cable features are easily visible. The cable is disconnected at the next substation at 292 m (*dashed cursor*). This allows calibrating distances. *Solid cursor* is at joint at 199 m (93.5 m back). Features on the *right side* of the open cable end are the same reflections just bouncing back and forth (distances are all multiples)

TDR performs best on clear cable breaks and low-ohmic short circuits between conductor and earth. High-ohmic faults are more difficult to detect. These faults can be burnt down with special "thumpers", which send high-energetic pulses into the cable. The impulses discharge through the fault, increase the devastation and decrease the fault resistance for better fault detection. The possible fault detection distance is depending on the cable design and instrument characteristics. Sharp events like the far end of the cable can be detected over hundreds of kilometres.

The example in Fig. 8.7 is taken from a land cable TDR measurement to illustrate some TDR features. Submarine cable measurements, especially on long length, are more obscured. Multiple reflexions can occur as well as ghost features of unknown origin.

### *8.3.2 Bridge Measurements*

Another useful fault location principle is based on resistance measurements in the conductor from one cable end to the fault. Using variations of resistance bridges, many different bridge measurement schemes are known since the 1880s [20]. A bridge measurement can be done if there is a healthy return conductor from the distant cable link station to the station where the measurement is performed. This second conductor forms one of the legs of a bridge measurement and must have a well-known resistance. In case of a three-phase submarine cable, another conductor in the same three-phase cable can be used if it is healthy. Also, a parallel cable can be used for the bridge measurement.

Figure 8.8 shows the Murray scheme for a bridge cable fault location. The cable has the length $L$ and the distance to the fault is $x$. At $R_f$ the insulation fault has contact to ground or seawater. For the purpose of measurement, the faulty cable core is connected to a healthy cable core in the same cable or a parallel cable at the far end (right end in the diagram). A battery voltage is fed to the circuit. The two tuneable resistances $R1$ and $R2$ are adjusted until the voltage meter $G$ shows zero ($G$ stands for "galvanometer" as these things were used in the 1880s to perform this

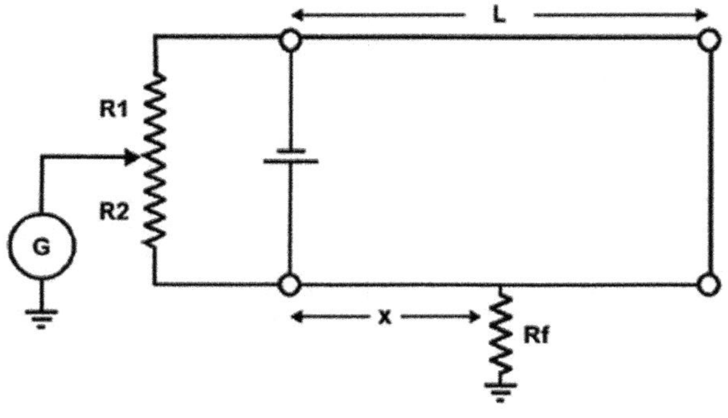


**Fig. 8.8** Murray loop for bridge measurement of cable fault location. Explanations to be found in the text

type of measurements). When $G$ is balanced to zero, the distance $x$ to the cable fault can be determined as

$$x = \frac{R_2}{R_1 + R_2} \cdot 2L \tag{8.3}$$

The Murray test loop is essentially an application of the Wheatstone bridge known from secondary school. Many other bridge applications have been developed within 150 years of fault location. Today, bridge measurement equipment for fault location with a high degree of operation convenience is commercially available. The bridge measurement instruments usually display the fault location in percent of the distance from the measurement place to the far-end station. Actually, this displayed value is the percentage of resistance-weighted distance. The displayed result must be compensated for possible conductor resistance differences along the route. These differences can be caused by differences in conductor size or temperature *en route*. Also, the resistance of the jumper cable between the power cable terminations at the far end has to be taken into account. This is a neat calculation task for the junior cable engineer. The bridge measurement is simple and can localise a fault within 0.5–1% of the cable length, which is good enough to start planning the repair work.

### *8.3.3 Fine Localisation*

The methods mentioned above are able to find the fault within a few percents of the total cable length. This can still be a few kilometres, too much for excavating the cable and searching with a ROV for the fault. Instead, there must be a fine localisation to pin-point the failure.

The fine localisation cannot be done from shore alone but requires an offshore visit to the suspected area. Fortunately, the localisation equipment is rather small and can be carried by small vessels.

With all modern fine localisation, a signal current is sent into the conductor of the faulty cable phase from shore. At the cable fault, the signal current exits through the insulation damage, which creates a difference of magnetic field on both sides of the fault location. A search coil on-board of the search vessel records the characteristics of the magnetic field from the signal current. Passing over the fault location, the search coil records a significant change in signal intensity. For a search coil on-board, the accuracy is about twice the water depth, provided a straight unburied cable.

In calm shallow waters, the search coil can be suspended from something like a fishing boat. More often, this is not enough. The search coil must be lowered towards the sea floor, requiring a winch. In many cases, an ROV is very useful to position a search coil closer to the cable, especially when the cable is buried. The use of an ROV requires vessels with specially trained teams and hoisting facilities. ROV-mounted cable-tracking equipment is very useful to locate buried cables.

Signals for fine localisation must be provided from land-based signal generators. Sinusoidal currents can be impressed onto the cable conductor. To avoid disturbances from the vessel's own power system the signal frequency should be different from 50 and 60 Hz. Sometimes very good results can be achieved when an impulse generator is used sending hv impulses into the faulty cable. These impulses can be "heard" by the search coil travelling along the submarine cable. Again, at the fault location the registered signal change its characteristics. Depending on the circumstances the fault can be found within a hundred meters. An ROV equipped with a search coil and which is flying close above the cable has a good chance of tracking down the fault location to a very short range.

If the fault location is located within a range of a few hundred meters, another method can be used. The cable is cut at the suspected damage location or in the middle of the suspected area. Both ends can be retrieved (one by one) to the vessel and a TDR measurement can be performed into each end. As the fault must be pretty close to one end, it will be easily detected and localised by the TDR.

### *8.3.4 Optical Time Domain Reflectrometry*

If the damaged submarine power cable contains optical fibres, OTDR (Optical Time Domain Reflectometry) can be used to localise damages of the fibres. Modern OTDR methods are well developed from telecom cable applications and are easy to use. Similar to TDR, OTDR is based on pulse travelling in the optical fibre. Fibre defects scatter a portion of the impulse light back to the monitoring unit, where the engineer can detect the fault location. Some special characteristics of OTDR methods should be kept in mind.

Most OTDR instrument cannot monitor the first hundred or so metres of optical fibre next to the optical unit. This area is called the blind spot. To compensate for this, OTDR engineers connect an optical launch cable (up to a thousand metres) between the instrument and the optical fibre in the tested cable. The length of the launch cable must be taken into account for the length evaluation.

The impulse travel velocity in the fibre is $V = c/n$ where $c$ is the speed of light in vacuum (300,000 km/s) and $n$ the refraction index of the glass fibre. It is obvious that the exact knowledge of the fibre refractive index is a prerequisite for good accuracy. Table 8.3 gives approximate values for the refractive index $n$ of fibres [21]. The fibre cable manufacturer should be able to produce exact values.

The OTDR method can only detect optical distances within the fibre. To translate the evaluated distances to power cable length, it is necessary to take into account

**Table 8.3** Refractive indices of optical fibres to be used in OTDR measurements

| Fibre type | Wavelength (nm) | Refractive index $n$ |
|---|---|---|
| Multimode | 850 | 1.481–1.495 |
| Multimode | 1200 | 1.476–1.490 |
| Single-mode | 1310/1550 | 1.467 |

1. the excess length of fibre within the optical cable
2. the excess length of the optical cable within the power cable due to the lay length.

The optical cable manufacturer should be able to specify the excess length of the fibre within the optical cable which is in the range of 0.05–1%. Then the optical fibre is incorporated into the power cable with a certain lay length. The lay length determines the fibre length in relation to the power cable length. Usually, the optical fibre is 2–8% longer than the power cable.

### *8.3.5 Other Methods*

If the damaged cable connects two transmission substations, the first information about the possible location of a cable fault can sometimes be obtained from the fault recorders. A precondition for this is that both stations have recorded the voltage shape during the breakdown and used exact time stamps on the voltage records. This is often the case in modern HVDC converter stations. The voltage breakdown at the fault location travels towards the cable ends with the wave group velocity, which in a mass-impregnated cable is around 150 m/μs. A rough estimation of the fault location can be obtained from the time difference between the arrivals of the breakdown signal at the two stations. The group velocity of the actual cable can be obtained by methods described in the TDR chapter.

Recently, a new cable fault location method has been suggested depending on the measurement of the complex cable impedance for different frequencies [22]. The group, which developed the method, claims that it also can detect global or local cable insulation degradation as there is, e.g., ageing. However, it is not known if it has yet been used for successful cable fault detection.

A rather unwanted fault location method has been used when searching for faults in oil-filled submarine cables.

The oil slick from the damaged leaking oil cable can sometimes be observed at the surface over the cable damage. At calm weather, this observation can be made from the air with a good chance to find the fault [17]. This method helped to localise a fault in the 400 kV a.c. cable between Denmark and Sweden in August 1979. Also after the damages striking the a.c. cable between the islands of Leyte and Luzon (the Philippines) the oil slick could be observed from the air.

Still another fault locating method is using the galvanic effect of a d.c. current. When a strong d.c. current is sent into the damaged cable core, a galvanic reaction can occur at the fault location, which makes gas bubbling into the water just above the fault. The gas development can be observed with ROV cameras.

A quite simple, still effective method was used to find some of the many cable faults in the notorious Fox Island cables. As the cable route was not known exactly, TDR or a similar method would have revealed the fault distance to land but not it's exact position. Instead, the repair vessel picked up the cable at shore and let it travel over wheels in the bow and aft. The vessel underruns the cable while travelling out to sea. Guided by the emerging cable, the vessel progressed with on-board specialists who searched the cable for the breakdown [22]. This method has been used already in 1853 when a telegraph cable off the Netherlands had to be repaired [23].

## 8.4 Repair Example

The following example illustrates step by step how a submarine cable repair operation can be done. It should be kept in mind that the described case is just one possibility, taken from a real situation. For each task, the experienced installation engineer will devise a tailor-made repair scheme, taking into account the particular conditions of the cable, the equipment, the location, and the weather.

The damaged cable runs from east to west at 50 m water depth. A flexible repair joint is available for the cable type. The crew of the fault location vessel has cut out the damaged portion (about 200 m of cable). The cable ends have been sealed properly and deployed on the sea floor. The cable ends have been provided with transponders, a ground wire, and rings and hooks to enable retrieving with an ROV.

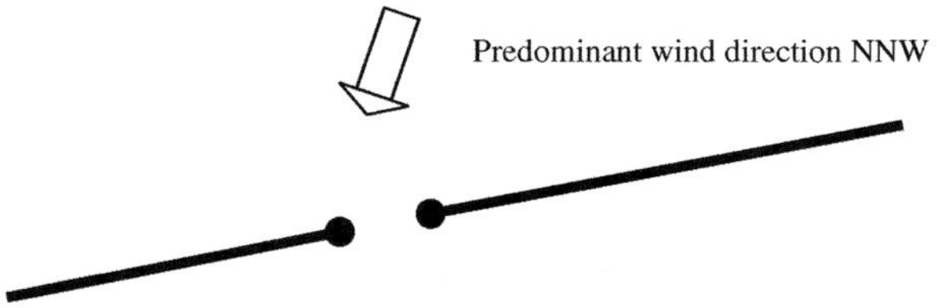


The repair vessel arrives on site with about 1000 m of spare cable on the turntable. Now it is important to plan the repair works so that the repair vessel will head into the wind during the delicate jointing work as much as possible to attain the largest possible stability.

The next sketch describes the repair vessel arrangement.

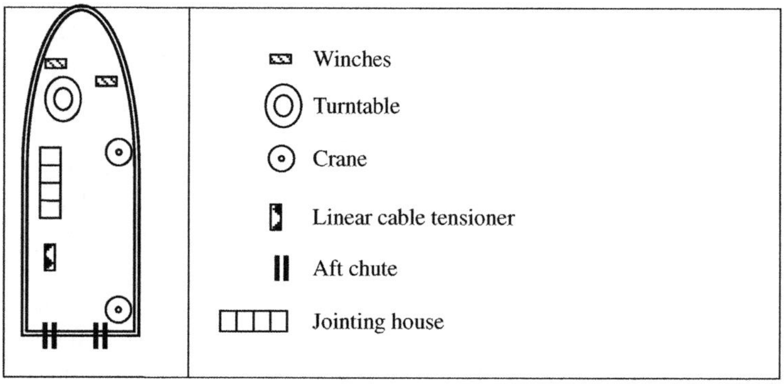

1. The east cable end is retrieved by ROV and pulled on board via the port side winch[3] over the port side aft chute. About 300 m of the cable are retrieved and stored on the turntable on top of the spare cable.

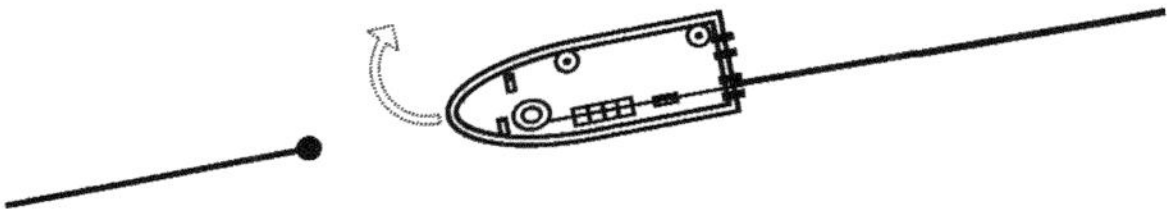

2. The repair vessel is moving in a curve towards a northerly direction, while the retrieved cable is laid out again over the port side chute. About 200 m of cable are now laid on the sea floor heading towards NNW. Transponders, ground wire, and ROV-friendly grip rings are attached to the cable end before it is put back down.

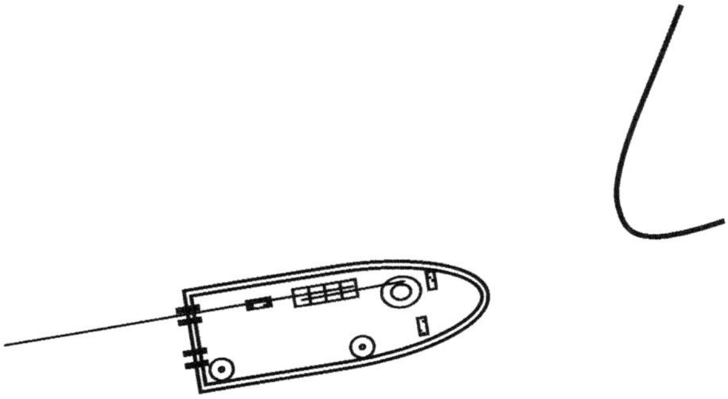

3. After laying down the end of the east cable towards NNW, the repair vessel is positioning itself so that the port side chute stands over the end of the west cable. The west cable is connected to a pull wire from the port side winch. The west cable is pulled onboard until the cable end is in the joining house. During the pulling the repair vessel moves backwards. The west cable is fixed aft of the jointing house by lashing or "chinese fingers". The spare cable is now pulled from the turntable into the jointing house.
4. The cable ends (west cable and spare cable) are now placed in the jointing house with overlap. Both ends are inspected with respect to damages or water ingress. If necessary, damaged pieces can be cut away and new cable can be pulled in from both the turntable or the sea floor. After the approval of the cable ends, they can be jointed in the jointing house. During this operation, the repair vessel can head up into the wind if necessary.

[3]Port side is the left side of the vessel when looking towards bow (front end). Starbord is the right side. The use of these terms is practical as they are non-ambiguous regardless of the ship direction on situation plans.

5. Once the joint is manufactured, the repair vessel is moving forward while paying out the flexible repair joint over the aft chute. The spare cable follows. When the joint and a short length of the spare cable have been laid safely onto the sea floor, the repair vessel is heading up in parallel to the end of the east cable, with a lateral distance of about 10 m. To reduce strain and forces on the flexible joint, it is advisable not to head up before the joint is in position on the sea floor. The two cable ends are now in parallel with a 10 m distance to each other.

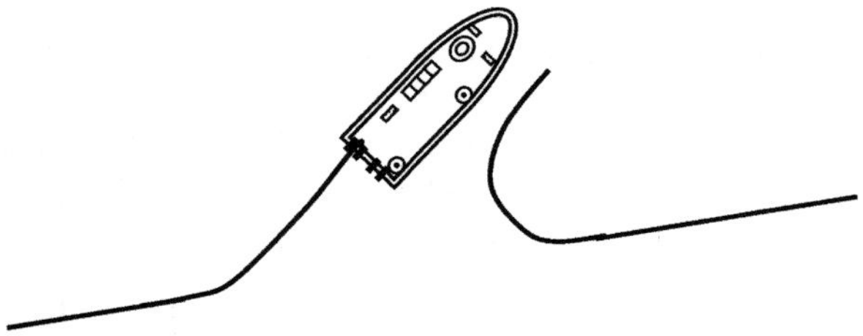

6. After passing by the end of the east cable on the sea floor, the repair vessel stops to pick up the end of the east cable, placing it over the starboard aft chute. Moving backwards, both cable ends are pulled back on board. On the port side chute, the spare cable is being pulled back onto the repair vessel and spooled onto the turntable. On the starboard chute, the east cable is being pulled up onto the vessel towards the vessel's bow.

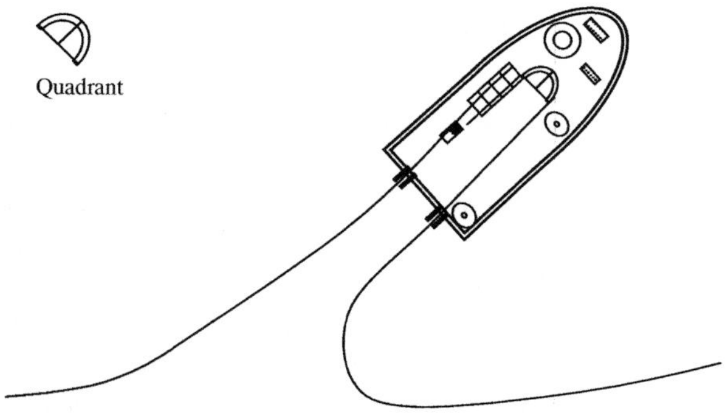


Now the quadrant is being introduced (cf. sketch, and Fig. 7.19 in Chap. 7). The 180-degree steel frame can hold half a turn of the cable and will later be used to deploy the cable into the water. The sketch shows the position of the quadrant on the vessel. While the vessel moves backwards taking in more cable, the eastern cable is being put around the quadrant and pulled into the jointing house until it lies there overlapping with the spare cable. The repair vessel is now positioned in the predominant wind direction and provides the best possible stability for the next jointing operation. During the jointing operation, the cables on the chutes must always be kept under tension so that the cables do not depart vertically from the chutes.

7. After jointing the cable link is restored, and can be tested from the cable ends on shore or on platforms if so desired. The cable loop is still on the repair vessel kept in place by the quadrant. Now, the jointing house must be removed. While the repair vessel is moving head-on slowly, the quadrant and cable loop are moved backwards on the vessel deck. The quadrant can be supported by rails, cranes, slipways or similar. The quadrant must be stabilized against tilting or warping. When the quadrant arrives at the aft chutes, it must be lifted carefully over the chutes by a crane (cf. Fig. 7.19). During this process, the repair vessel must maintain the cable tension by travelling forward. The quadrant goes down in the water suspended from the aft crane or the A-frame. As the repair vessel slowly travels ahead, the quadrant descends into the water and finally lands on the sea floor. An ROV can now unlash the cable from the quadrant. The repair is finished, and the cable can be protected. Depending on the joint manufacturing time and other circumstances, the described procedure can take 4–12 days.

## References


1. Lawrence M (2002). Submarine Cable Industry, -Alcatel Survey Standards and Future Trends. MUC Kiel, 2002.
2. Hoshina R (2001). Improvements in Submarine Cable System Protection, SubOptic 2001.
3. Shapiro S, Murray J, et al. (1997). Threats to Submarine Cable. SubOptic'97, San Francisco, USA.
4. Methods to Prevent Mechanical Damage to Submarine Cables, Presented by Cigré Working group 21 as Session Paper 21-12 at the 1986 Cigré Session, Paris, France.
5. Drew S and Hopper A (1996). Fishing and Submarine Cables Work Together, Pamphlet Issued by the International Cable Protection Committee, London.
6. Update of Service Experience of hv Underground and Submarine Cable Systems, to be Presented by Cigré Working Group B1-10 2009.
7. Nishimoto T (1997). Chugoku Goes Fishing For Submarine Cable Solution. Transmission and Distribution World, October 1997.
8. Christensen C F (2006). Cable Risk Assessment. Lecture on a Symposium on Installation of Submarine Power Cables for Grid Connection of Offshore Windparks. Bremen, Germany, 2006. Arrangend by DENA. http://www.offshore-wind.de/page/index.php?id=2609
9. Luger D (2006). Developments in Anchor Technology and Anchor Penetration in the Seabed. Lecture on a Symposium on Installation of Submarine Power Cables for Grid Connection of Offshore Windparks. Bremen, Germany, 2006. Arrangend by DENA. http://www.offshore-wind.de/page/index.php?id=2609
10. Nakamura M et al. (1992). Reliability Analysis of Submarine Power Cables and Determination of External Mechanical Protections, IEEE Transactions on Power Delivery, Vol. 7, No 2, pp. 895–902, April 1992.
11. Det Norske Veritas (2008). Environmental Conditions and Environmental Loads. Recommended Practice DNV-RP-C205, Norway, April 2007, Amended April 2008.
12. Geijer G et al. (1958). The Öresund Power Cable Project, Blue-White Series, Published by the Swedish State Power Board, 1958.
13. Article in the Massena Observer, 21st November 1974, Massena, New York, USA.
14. Houtz R E et al. (1962). Turbidity Current at Kadavu Passage, Fiji, Geological Magazine, Vol. 99, No. 1, pp. 57–62.
15. Milkov A V (2000). Gas Hydrates at Minimum Stability Water Depths in the Gulf of Mexico: Significance to Geohazard Assessment, Transactions of Gulf Coast Association of Geological Societies, Vol. L.

16. Chamberlin D M et al. (1979). The Long Island Sound Submarine Cable Interconnection Operating Experience, 7th IEEE/PES Transmission and Distribution Conference and Exposition, April 1–6, 1979.
17. Elgh L et al. (1982). Repair of 400 kV ac Submarine Cable. Paper 21-05, Cigré Session 1982.
18. Michel R et al. (1985). Les câbles terrestres et sous-marins et les travaux en mer, Revue générale de l'électricité, No. 2, 1985 (in French).
19. Lecture by Gord Parker, Radiodetection, Canada, http://www.radiodetection.ca/docs/imsa2001.pdf
20. Eberlein D (2006). Messtechnik Fiber Optic, Gemeinschaftsseminar, ISBN 3-00-018278-0
21. Fantoni P F (2006). Wire System Aging Assessment and Condition Monitoring Using Line Resonance Analysis (LIRA). 5th American Nuclear Society International Topical Meeting on Nuclear Plant Instrumentation, Controls, and Human Machine Interface Technology, November 12–16, 2006, Albuquerque, NM.
22. Private Communication by Wally MacDonald. http://marencoengineering.com/case_study.html
23. Wilkinson H D (1898). Submarine Cable Laying and Repairing, The Electrician Printing and Publishing Company, London.

# Chapter 9
# Operation and Maintenance: Reliability

## Contents



## 9.1 Operation of Submarine Cables

After commissioning, submarine power cables are expected to work troublefree for decades. The cable operator can contribute to a long useful life and a high reliability of the cables by some maintenance actions. This chapter explains what the cable operator can do to improve the availability.

### *9.1.1 Common Measures for All Kind of Submarine Power Cables*

The cable operator must protect the cable from a number of hardships, such as overvoltage, overheating, external violence, fatigue, etc. This requires an active involvement and the establishment of maintenance plans for the cable system. In most cases, a simple instrumental monitoring is sufficient.

Overvoltage can be avoided by a system analysis prior to system design. Surge arresters with appropriate protection levels should be used, especially when the

T. Worzyk, *Submarine Power Cables*, Power Systems,
DOI 10.1007/978-3-642-01270-9_9, © Springer-Verlag Berlin Heidelberg 2009

submarine cable is connected to overhead lines. Surge counters should be checked regularly in order to detect anomalies in the power system behavior. Temporary system overvoltage due to strong generation (e.g. in OWP) should be monitored carefully and mitigation strategies considered.

The risk of overheating can arise even if the cable is run within its design limits. Seafloor conditions can alter within months, and the cable route can be covered with additional thick layers of sediment. This reduces the heat flow from the cable, which leads to cable temperatures higher than expected. Also, other services such as pipelines, flow lines, other power cables, etc., may suddenly turn up in the vicinity of the power cable. The cable operator should develop strategies to detect this type of thermal hazards:

- Disseminating knowledge of the existence of the cable to authorities, infrastructure developers, utilities etc.
- Watch out for construction activities at the cable sites
- Study bathymetric updates from survey companies, marine authorities and professional associations
- Check and compare satellite photographs in regular intervals.

The thermal conditions along the cable route can be monitored with a DTS (Distributed Temperature measurement System). Adverse conditions can be detected, and relieving action can be launched.

Free spans in the cable route should be checked from time to time to detect a possible enlargement. Too long free spans can easily cause vibration, chafing, fatigue, and cable failures. Free spans can be stabilized with additional rock-dumping, or with mid-span guy lines. Another possibility to improve the vibration characteristics of free spans is the subsequent application of VIV suppression strakes.

A number of active measures can be taken to avert external damage to the cable. In Chap. 7, measures for the active post-lay cable protection are described. They range from information campaigns for the fishing community to repeated share of cable position data to authorities, nautical agencies, etc. Active ship traffic monitoring allows for warning of vessels, should these come too close or try to anchor too close to the cable route.

The cable operator can monitor the character and frequency of vessel traffic and fishing activities along the route of the cable. If the vessel traffic increases in size or numbers, the cable operator may consult marine authorities in order to discuss methods to divert traffic from the area.

### *9.1.2 Instrumentation*

Taking active control of the operation of a submarine power cable may increase the availability and profitability of the cable link. Suitable instrumentation is available to do this.

#### 9.1.2.1 DTS

The Distributed Temperature Measurement System (DTS) is an optical fibre-based temperature sensor incorporated into the power cable, or installed alongside the power cable. Such a system is able to monitor the temperature along the power cable. A land based monitoring unit evaluates the optical signals and displays a temperature profile over the cable length.

Hot-spots and cold-spots, and other irregularities along the cable route can be detected with DTS. By comparing DTS printouts from subsequent years, it is possible to detect sediment wash-away at the particular location. If a submarine cable is exposed in free spans or by wash-away of sediment it should be considered to take advantage of the situation and fly an ROV along the cable to inspect it for damages and corrosion. Submarine power cables that are not buried should be inspected regularly by divers or ROV to confirm their status or changes in their environment.

Most DTS have built-in ODTR capabilities to detect and localize fibre damage. But DTS can be used even if everything is running smoothly.

The temperature profiles recorded from this system can be evaluated to perform a dynamic cable rating. In contrast to the conventional static "Rated Power", the system is able to calculate the ampacity of the link by taking into account the actual ambient conditions. The cable operator can take advantage of the ampacity changes over the seasons of the year. Furthermore, dynamic rating can exploit thermal reserves in the cable system for temporary overloads. The operator of links designed for the purpose of power trading may use this system to generate extra income without putting the cable integrity at risk. Dynamic rating systems are also known under the acronyms CLPS or RTTR.

If the power cable is equipped with optical fibres for data transmission, these can be used for ad-hoc distributed temperature measurements.

#### 9.1.2.2 CDVC

The Cable Dependent Voltage Control, designed for mass-impregnated d.c. cables, is a control function resident in the HVDC converter station. At reduction of power demand, the system reduces temporarily the system voltage without limiting the transmission capacity. Doing so, the electric stress in the cable insulation is reduced when the cable is cooling down. The voltage reduction has the potential to increase the electric life of the cable. When full power is required again, the voltage returns to rated.

#### 9.1.2.3 Partial Discharge Monitoring

Partial discharge (PD) monitoring of power cable systems has achieved a sophisticated level. Defects, inhomogenities and other flaws can be evaluated with on-line PD detection systems. PD activities on a particular spot in the cable link may indicate the presence of a small damage in the cable insulation system or in a cable joint.

Modern PD measurement systems are getting increasingly powerful in discrimination of PD signals and their relevance for the future cable life. Still, it is extremely difficult to monitor a longer stretch of submarine power cables from the shoreline as PD signals tend to be attenuated after only a few kilometres of travel through the cable.

Furthermore, there is the great question about what happens if the system detects an increased PD activity in a particular joint or cable location out there. Would the TSO manager decide the interruption of the cable link for repair of a suspicious cable portion? Or would he/she prepare a repair operation but wait for the breakdown that may not happen?

### *9.1.3 Mass-Impregnated Cables and XLPE Cables*

Mass-impregnated and extruded cables, also called solid cables (cf. Chap. 2), are maintenance-free. This is also valid for the submarine joints belonging to these cable types. Cables with solid insulation have no free oil volume that needs to be pressurized from the shore stations. The terminations require only a small pressure vessel to accommodate the oil expansion inside the termination. The monitoring sensors for pressure and oil level can be connected to the substation SCADA system. Too low or high pressure would cause a tripping of the link.

XLPE insulated a.c. cables are sometimes terminated directly into GIS substations. Not all, but some GIS cable terminations have small amounts of insulation oil (mineral oil, polyisobutene oil or silicone oil) calling for level/pressure monitoring. The maintenance requirements are identical to those applicable for equivalent terminations for land cables.

### *9.1.4 LPOF, SCOF and SCFF Cables*

Fluid-filled (FF) and oil-filled (OF) cables require the monitoring of the oil-pressure feeding system. Since the performance of these cables is critically depending on the prevailing oil pressure, it must be monitored on-line. A loss of oil pressure indicates a cable damage calling for immediate attention. The pressure monitoring can be connected to the substation SCADA system. The cable link will be tripped if the pressure falls under a certain limit.

Sufficient amounts of degassed dielectric fluid must be available in the feeding station in order to maintain a positive oil pressure also in the case of a cable leakage. As leaks may be large and weeks may elapse until the cable can be sealed, it should be considered to keep an extra stock of cable oil not too far from the termination. Cable operators of one region can cooperate in this issue, as cable oils from different cable systems usually are compatible.

### 9.1.5 Cable Terminations

The proximity of many submarine power cable terminations to the sea may cause severe salt layer deposits on the insulators calling for a regular cleaning schedule. Some cleaning can be done under energized lines. Some termination insulators rely on the hydrophobicity of polymeric materials or coatings. The remaining hydrophobicity should be tested regularly, e.g. according to visual tests suggested by STRI [1] and set forth in an IEC specification [2].

All terminations erected close to shore or on marine platforms must be checked regularly for corrosion. Especially galvanic corrosion between different metals can be onerous.

## 9.2 Reliability of Submarine Cables

Submarine power cables are often a special asset in the basket of TSOs and might deserve closer attention. The operating utilities are very much interested in a trouble-free operation because repair is expensive and can be cumbersome. Also, submarine power cables are often not a part of a meshed network with large redundancy. On land, the power flow blocked by a cable failure can most often be rerouted through other paths of the grid. In contrary, submarine power cables often have no redundant grid for back-up. A submarine cable failure can darken islands or oil/gas production platforms, or cut the revenues from offshore wind farms. A failure in submarine cables connecting self-sustaining grids (such as HVDC links between countries) would not black-out cities, but would deprive the owners of large revenues from power trading. For this reason, the reliability of submarine power cables is an important aspect of each business model and has strong influence on the cable design.

Unfortunately cable operators are somewhat reluctant about reporting failure statistics. A number of journal articles were published in the 1970s and 1980s reporting on experiences from large submarine cable links including accounts of failure modes and repair methods. In the last decade reports of this kind have become astonishingly scarce. Owing to the excellent work of the Cigré Study Group B1 (erstwhile SC 21), cable operators have the possibility to report cable experiences and failures confidentially without annoying the shareholders.

### 9.2.1 The Cigré Studies

The Cigré organisation (Cigré=Conseil International des Grands Réseaux Électriques) has established a working group that regularly collects data on submarine power cable reliability. In 1986 it published an often-cited compilation on experiences from thousands of kilometres of submarine power cables [3]. The

**Table 9.1** Failure rates for submarine power cables > 60 kV expressed in (failures/(year×100 circuit kilometres))

| | Internal origin failures | External origin failures | All failures |
|---|---|---|---|
| a.c. HPOF cables | 0 | 0.7954 | 0.7954 |
| a,c, LPOF cables | 0 | 0.1189 | 0.1189 |
| a.c XLPE cables | 0 | 0.0706 | 0.0706 |
| d.c. MI cables | 0 | 0.1114 | 0.1114 |
| d.c. LPOF cables | 0.0346 | 0 | 0.0346 |

study covers experience from 1950 to 1980. It arrives at a failure rate of 0.32 failures/year/100 cable kilometres. Only a small fraction of the cable kilometres considered in this failure rate are protected. Today, since the protection rate of submarine power cables is considerably higher, and survey methods, cable design, and installation methods have developed enormously, the failure rate is expected to be much lower.

According to the 1986 Cigré study 82% of the failures occurred in the cables and 18% in the joints. The dominant majority of the cable faults were caused by external violence, while the joint failures mostly were caused by poor engineering, installation or maintenance.

The failure rates reported in the 2009 Cigré study [4] are summarized in Table 9.1. Except for HPOF submarine cables (which are used very rarely today), all failure rates are much lower than the 1986 values. The values are related to circuit kilometres. As circuit often comprises two, three or four individual cables, the rate as per cable kilometre would be even lower.

While the ratio of joint failures to cable failures was 0.22 in the 1986 study, this relation changed to 0.095 in the 2008 study. This indicates a much higher relative safety of cable joints in the recent time.

It should be noted that calculated failure rates can be warped by a few notorious cable links, which account for a significant share of the reported failures.

## *9.2.2 Failure Statistics for Large HVDC Cable Projects*

The failure rate calculated from the data given in Table 9.2 is 0.264 failures/year/100 cable kilometres for mechanical faults and 0.0143 failures/year/100 cable kilometres for other failures. "Other faults" means internal faults, such as the failure of the insulation system. It is evident that a few cable systems contribute largely to the failure statistics. In particular, the 1964–1988 Kontiskan 1 suffered many mechanical failures. Without the contribution of the Kontiskan 1 link the failure rate would be only 0.1 failures/year/100 cable kilometres. It is obvious that badly engineered cables or unsuitable installation methods account for the majority of cable failures.

**Table 9.2** Failure in some large HVDC cable links as per 1998[5]

| Cable project | V [kV] | P [MW] | L [km] | Max depth [m] | Cables [#] | In service year | Service time years | Technical service [km*years] | Mech faults [#] | Other faults [#] |
|---|---|---|---|---|---|---|---|---|---|---|
| Kontiskan 1 | 285 | 300 | 88 | 80 | 1 | 91 | 4 | 352 | 0 | 0 |
| Cook Strait | 350 | 185 | 41 | 260 | 3 | 91 | 4 | 492 | 0 | 0 |
| Fenno Skan | 400 | 500 | 200 | 120 | 1 | 89 | 6 | 1200 | 1 | 0 |
| Konstiskan 2 | 285 | 300 | 88 | 80 | 1 | 88 | 7 | 616 | 0 | 1 |
| Cross-Chn. 2 | 250 | 200 | 48 | 60 | 8 | 86 | 9 | 3456 | 0 | 0 |
| Hokkaido-Hon. | 250 | 200 | 43 | 550 | 2 | 78 | 17 | 1462 | 0 | 0 |
| Vancouver 2 | 280 | 100 | 32 | 200 | 2 | 76 | 19 | 1216 | 1 | 0 |
| Skagerrak $^1/_2$ | 250 | 250 | 128 | 550 | 2 | 76 | 19 | 4864 | 6 | 0 |
| Vancouver 1 | 260 | 100 | 32 | 200 | 3 | 69 | 26 | 2496 | 3 | 0 |
| Sardina-Cor. | 200 | 100 | 118 | 450 | 2 | 67 | 28 | 6608 | 12 | 0 |
| Kontiskan 1 | 285 | 250 | 87 | 80 | 1 | 64–88 | 24 | 2088 | 44 | 2 |
| Gotland 1 | 150 | 30 | 93 | 170 | 1 | 54–89 | 35 | 3162 | 7 | 1 |
| Total | | | 998 | | 27 | | 198 | 28012 | 74 | 4 |

### *9.2.3 Definition of Reliability Terms*

Reliability terms can be expressed in many different terms. For a correct comparison of information from different sources, it is important to use a well-defined terminology.

| | |
|---|---|
| Reliability | The probability that a cable is fulfilling its purpose adequately for the period of time intended. The reliability can be described as: $1-(\lambda \times r)/8760$ |
| Availability | A measure of a system's performance in terms of its reliability and maintainability: $1-(\lambda \times r + c)/8760$ |
| Where | $\lambda$ = number of failures per year<br>r = repair time after failure [hours]<br>c = scheduled outage (e.g. maintenance) [hours/year] |
| Outage | Period of non/functioning of the system |
| Forced outage | Involuntary outage as a result of a failure |
| Failure rate | The annual rate of forced outages associated with failures in the cable |
| Scheduled maintenance | Annual preventive maintenance as specified by supplier or according to operators own standards |
| Unscheduled maintenance | All maintenance required that can not be termed annual preventive maintenance |

### *9.2.4 Reliability of Some Specific Submarine Power Cables*

A few reports on reliability and availability data can be found in the literature. However, cable operators today are less prone to publish such data. This is a disadvantage to the industry, as investors and insurance companies might need a clearer picture for a fair risk assessment. Here, a few examples are given.

#### 9.2.4.1 Skagerrak HVDC Scheme

The first Skagerrak HVDC cable between Norway and Denmark was installed 1976–1977. The base data are as follows:

| | |
|---|---|
| Rated voltage | 250 kV d.c. |
| Number of cables | 2 |
| Year of commissioning | 1976, 1977 |
| Submarine distance | 128 km |
| Largest depth | 550 m |

The operation statistics for the interval between 1978 and 1984 is reported in [6]. For this pair of cables the availability was always better than 93% during this time span, despite forced outages due to cable failures in two of seven years. In five of seven years the availability was over 95%.

#### 9.2.4.2 Windfarm Export Cable

For a hypothetic OWP export cable the reliability of five different a.c. cable configurations is compared in [7]. Three-core and single-core systems have been investigated for a fictive 60 km 1000 MW power export system. A 275 kV single-core system with seven individual cables (two circuits and one spare cable) was found to have the lowest expected failure rate (0.15 failures per year and cable) (Table 9.3):

**Table 9.3** Failure rates for different cable schemes to connect a large OWP

| Voltage (kV)<br>cable type | 132<br>3-core | 220<br>3-core | 400<br>3-core | 275<br>SC | 400<br>SC |
|---|---|---|---|---|---|
| No of cables for 1000 MW transmission | 6 | 4 | 2 | 7 (1 spare) | 4 (1 spare) |
| Failures/year and cable | 0.25 | 0.46 | 0.67 | 0.15 | 0.22 |

Taken into account the number of cables, the 400 kV system with three single-core cables seems to have the lowest expected number of repairs over lifetime.

#### 9.2.4.3 Fox Islands

Four SC submarine cables were laid between Rockport, Maine, USA, and the Fox Islands in 1976. The 10.4 miles of 34.5 kV circuit replaced on-island diesel generation. The cable system was stricken by 45 faults until decommissioning in 2005 [8]. While the Cigré failure statistics for submarine power cables [1] quote a rate of 0.32 faults per year and per 100 km of cable system, the Fox Island cable counts nine faults per year and per 100 km, and has perhaps the worst reliability record in the world.

#### 9.2.4.4 Long Island

Seven single-core high-pressure oil-filled 138 kV cables were installed under the Long Island Sound and commissioned in 1969. At that time it was the longest oil-filled submarine cable in the world [9]. Eighty percent of the cable route was installed freely on the seafloor, with 275 m spacing between individual cables [10]. Unfortunately, the cable system (called "1385 cable") had an exceptionally bad availability. The operator Long Island Power Authority reports:

> Since being energized in 1969, the existing 1385 Cable had experienced 36 incidents with 58 damages to cables that have resulted in either limited capacity operation or total electrical failures. The causes for incidents included chafing from rock and corrosion, and bottom dragging of grappling hooks or anchors from both small and large vessels. As a result, frequent and extensive maintenance was required for the existing 1385 Cable. Since 1990 alone, the cost of cable repairs has exceeded $45 million.

In spite of the extra spare cable and the spaced installation multiple damages caused 50% capacity drop in extended periods (2–9 months) in the 1970s. The outages were caused mostly by external damages affecting a system already weakened by damages that were caused by severe corrosion. No less than four cables were damaged by a single anchor attack 17 November 2002.

The unlucky cable system is now replaced by a system of three 3C cables with 138 kV rating. Since the new cables are protected by burial six feet under the seafloor, they have much better chances to serve without trouble during many years.

#### 9.2.4.5 Baltic Cable

The Baltic Cable is a HVDC cable connecting Sweden and Germany. The return current is conducted by sea electrodes.

| | |
|---|---|
| Rated voltage | 450 kV d.c. |
| Number of cables | 1 |
| Year of commissioning | 1994 |
| Submarine distance | 250 km |
| Largest depth | 40 m |

The performance data of the Baltic Cable link are summarized in Table 9.4 (Courtesy of Baltic Cable AB, Sweden).

Three failures were reported, one in 1999, 2002 and 2009, resp. This figure calculates to 0.08 failures per year and 100 cable kilometres, which is better than the reported average in both the 1986 and 2009 Cigré studies. The 1999 cable failure led to 3374 h of outage[1] in 1999 and 2000 reducing the availability figure for 2000 drastically. The 2002 failure could be repaired within 8 weeks [11].

**Table 9.4** Failure rates for different cable schemes to connect a large OWP

| Year | Scheduled unavailability (%) | Forced unavailability (%) | Availability (%) |
|---|---|---|---|
| 1995 | 3.14 | 0.82 | 96.04 |
| 1996 | 6.00 | 1.24 | 92.76 |
| 1997 | 1.49 | 1.40 | 97.11 |
| 1998 | 1.50 | 2.29 | 96.21 |
| 1999 | 1.29 | 7.67 | 91.04 |
| 2000 | 1.23 | 30.76 | 68.01 |
| 2001 | 1.23 | 0.78 | 97.99 |
| 2002 | 0.96 | 14.08 | 84.96 |
| 2003 | 1.23 | 0.01 | 98.76 |
| 2004 | 8.88 | 0.04 | 91.08 |
| 2005 | 1.23 | 0.61 | 98.16 |
| 2006 | 1.23 | 1.52 | 97.24 |
| 2007 | 1.23 | 0.14 | 98.63 |

[1] The cable repair was delayed by vessel unavailability, unsuitable weather and other factors.

# References


1. Gutman I (2004). STRI Guidelines for Diagnostics of Composite Insulators. Proceedings of International Conference on Suspension and Post Composite Insulators: Manufacturing, Technical Requirements, Test Methods, Service Experience, Diagnostics, St. Petersburg 4–9 October 2004. See also www.stri.se
2. IEC Technical Specification TS 62073: Guidance on the Measurement of Wettability of Insulator Surfaces. Published by IEC (International Electrotechnical Commission) in 2003.
3. The CIGRE Paper (1986). Methods to Prevent Mechanical Damage to Submarine Cables, Presented by Cigré Working Group 21 as Session Paper 21-12 at the 1986 Cigré Session, Paris, France.
4. Cigré Study Committee B1 Insulated Cables (2009). Update of Service Experience of HV Underground and Submarine Cable Systems, to be presented by Cigré Working group B1-10 2009.
5. Reliability, Availability and Maintainability Study NorNed HVDC Kable Project, 1998 http://www.dte.nl/images/12_24746_tcm7-4788.pdf, and http://www.tennet.nl/images/12-B2-Berdal_Stromme_RAM_study_tcm41-12291.pdf
6. Hauge O et al. (1988). Performance of the ± 250 kV HVDC Skagerrak Submarine Cables, IEEE Transactions on Power Delivery, Vol. 3, No. 1, pp. 1–15, January 1988.
7. Svoma R et al. (2007). Subsea Connections to High-Capacity Offshore Windfarms, Paper A 9.5, Jicable 2007, Paris, France.
8. Private Communication by Wally MacDonald. See also: http://marencoengineering.com/case_study.html
9. Chamberlin D et al. (1797). The Long Island Submarine Cable Interconnection, 7th IEEE/PES Transmission and Distribution Conference and Exposition, April 1976.
10. Private Communication Long Island Power Authority, LIPA
11. Private Communication Baltic Cable AB, Sweden

# Chapter 10
# Environmental Issues

## Contents



## 10.1 Environmental Assessment

For all industrial projects, the environmental impacts must be considered, the relevance of these impacts must be evaluated, and possibilities to reduce these impacts must be discussed. The delicate state of our planet, the industry's responsibility for its activities, and the public attention require a serious and honest assessment. Those who want to erect an industrial project such as a submarine power cable must perform regulated Environmental Impact Assessments (EIA). Within the EU, the directives 97/11/EG and 85/337/EEG regulate the legal background for the EIA process. Furthermore, authorities call for compliance with laws and rules.

The installation of a submarine power cable requires permits to be obtained in sometimes lengthy processes. Steps to be taken, and authorities to be approached, differ from country to country, and are subject to legislation, that sometimes is volatile. A treatment of permit issues would be insufficient in the framework of this book. Instead, some basic technical issues on the environmental aspects of

T. Worzyk, *Submarine Power Cables*, Power Systems,
DOI 10.1007/978-3-642-01270-9_10, © Springer-Verlag Berlin Heidelberg 2009 

submarine power cables are discussed here in order to facilitate a fact-based fruitful discussion between project developers, authorities, and the public.

The environmental impacts of submarine power cables have to be considered "from cradle to grave", i.e. all aspects of raw material production, transports, manufacturing, installation, operation, recovery, and recycling add to the overall impact of a submarine cable. Various LCA (Life Cycle Assessment) models are available. LCA models work with various categories of environmental impact. For a given product, the LCA can compute and compile the aggregate environmental impacts starting from the needed amounts of constituent and auxiliary materials, consumables, and energy. LCAs do not create new data; they just compile appropriate data from the underlying database. The quality of an LCA is hence depending on the quality of the data base entries. As an example, for the assessment of the influence of the use of copper, it is essential to make correct assumptions on the share of recycled and virgin material, as well as the mining and refining site on the globe. The skills of the LCA engineer are critical to assemble a correct model of the product and its peripheral processes, and to interpret the results. Various LCA tools are available on a commercial basis. The EIME tool has been developed by 6 major electronic companies 10 years ago, and has been adopted by "la Fédération des Industries Electriques, Electroniques et de Communication" (FIEEC) [1]. This tool has the following categories of environmental impact:

| | |
|---|---|
| RMD | Raw material depletion |
| ED | Energy depletion |
| WD | Water depletion |
| GW | Global warming |
| OD | Ozone depletion |
| AT | Air toxicity |
| POC | Photochemical ozone creation |
| AA | Air acidification |
| WT | Water toxicity |
| WE | Water eutrophication (too much nutrients in water bodies) |

Other tools have also Hazardous Waste Production as a category.

LCAs can support decision making on submarine power cables in different levels:

1. On the lowest level, the design engineer may apply LCA on different cable design solutions meeting the functional specification of the submarine power cable. Based on the result, the engineer may select the environmentally best solution.
2. If the purpose of a submarine power cable is to transmit power between two given points A and B, the LCA can help to rank possible routes with respect to their environmental impact. Alternative routes can sometimes also include overhead lines or land cables. In this case also demographic and socio-economic factors need to be taken into consideration.

3. Sometimes permission authorities require an environmental comparison between the planned project and the "zero alternative", i.e. the situation when nothing is being built. For submarine power cables, the zero alternatives could be continued operation of Diesel generators on unconnected islands, or the need of costly spinning reserve in autonomous grids. For wind farm export cables, the zero alternative would be that the harvested wind energy would not be available and must be produced in a different way, possibly with fossil fuels.

Unfortunately, LCA results are sometimes being used by various parties to promote their own agenda. Green groups, NGOs, and industrial lobbyist groups work with LCAs to proof their statements. Also, political parties, companies, and investment funds promoting "green" investments often present fragmentary LCA graphs or tables.

The task of this chapter is not to present the results of complete LCAs. This should be performed using individual data for each project. Here, we want to discuss some features specific to submarine power cables, which may help the LCA analyst to fine-tune models and assumptions. The factors related to the production and transport of raw materials are not listed here because they are highly depending on specific project circumstances, and would be outdated after only a few years. Commercial LCA tools hopefully provide updated databases.

## 10.2 The Influence of Cable Losses

A standard LCA includes the impact from the operation of a product/asset. The transmission losses of a power cable during service generate a much higher environmental footprint than the energy used for raw material production and manufacturing. In fact, the transmission losses and the environmental impact from their generation obscure almost all environmental impacts from raw material production, manufacturing and transport of the submarine power cable [2]. This leads to the unfortunate situation that the cable design engineer can put almost any questionable material in the cable without affecting the cradle-to-grave analysis. Only if the environmental impact from the cable operation losses is excluded for the moment, the environmental difference between design solutions becomes visible.

The environmental impact of transmission losses can only be assessed correctly with the knowledge of how the power for these losses is generated. Even if all facts and figures are firmly available, the conclusion is far from being unambiguous. As an example, the impact of transmission losses from an OWP submarine cable can be evaluated in at least four different manners:

1. The transmission losses are generated by clean wind power, hence the impact is zero.
2. The losses eat up some of the wind power arriving onshore and decrease the possibility of replacing fossil power generation. The impact of losses is the same as of an equivalent amount of fossil-generated power.

3. The wind power does not replace fossil power alone, but a typical power-mix in the country where the wind power is landed. This can be fossil, nuclear and sustainable power.
4. The losses are compared to an international power-mix of the future.

Table 10.1 illustrates the difference in $CO_2$ emission per kWh depending on the choice of reference power production [3]:

**Table 10.1** $CO_2$ emission per kWh for different system borders

| System borders for electrical power production | Average $CO_2$ emission Kg $CO_2$/kWh (2005) |
|---|---|
| Sweden | 10 |
| Scandinavia | 58 |
| EU (25) | 415 |
| Marginal power production from coal-fired plants | 400–750 |

In any case, the reduction of losses improves the environmental impact of a submarine power cable more than any other detail.

## 10.3 Environmental Aspects Related to Cable Design

### *10.3.1 Conductor Materials*

Within the framework of the project specification, the design engineer often has little possibilities to choose materials from an ecologic point of view. Submarine power cables can be designed with aluminium or copper conductor. The conductivity of copper is by 64% better than that of aluminium.

Table 10.2 shows the required amount of energy for the production of copper and aluminium conductors with equal conductance. It is obvious that a larger cross section is necessary for the aluminium conductor to achieve the same conductance, but a lower mass because of the low density of aluminium. With the base data and method used here, the energy required for the production of equivalent copper or aluminium conductors is almost equal.

A cable with aluminium conductor, however, requires more insulation and armoring material due to the slightly larger conductor diameter.

### *10.3.2 Choice of Other Cable Materials*

The choice of insulation material is determined by the intended use of the cable and the state-of-the-art. Only sometimes there is a free choice.

Most modern submarine power cables have an insulation made from XLPE. The peroxide-based crosslinking process in the factory releases gaseous by-products such as acetophenone, alpha-methylene-styrene and cumyl-alcohol. These exhaust

**Table 10.2** Energy vs. conductance needed to produce conductor material

| | Cu-conductor, 1200 mm$^2$ | Al-conductor, 2000 mm$^2$ |
|---|---|---|
| Conductor resistance at 20°C | 0.015 mΩ/m | 0.015 mΩ/m |
| Conductance | 66.666 m/Ω | 66.666 m/Ω |
| Mass of conductor | 10.7 kg/m | 5.4 kg/m |
| Energy used for raw material production [4] | | |
| Primary raw material (freshly mined) | 60 GJ/t | 164 GJ/t |
| Recycled raw material | 20 GJ/t | 20 GJ/t |
| Share of recycled material $\alpha$ | 0.9 | 0.8 |
| Energy used to produce raw material from mix virgin/recycled | 24 GJ/t | 48.8 GJ/t |
| Energy used for raw material for 1 m of conductor (MJ/m) with equal conductance | 256.8 | 263.5 |

gases from the manufacturing are taken care of in the factory according to local regulations but might contribute to air toxicity. Later, during operation, no more by-products will be released into the atmosphere or water.

XLPE cannot be recycled by melting to new products. Instead, its energetic content can be used in suitable incineration plants releasing water and carbon dioxide. XLPE does not contain halogenides. However, some additives are used in the compounding of XLPE resin to improve production and electrical properties.

Thermoplastic polyethylene is an alternative extruded insulation material. The absence of cross-linking has some important environmental advantages. First, cross-linking by-products are not being generated. Second, less process energy is used in manufacturing. Third, thermoplastic polyethylene can be recycled easily.

The semiconducting layers of the conductor and insulation screen are made from carbon-black filled polymers that are compatible to the insulation material. Due to the cross-linking and the inseparable carbon-black filling these materials can hardly be recycled. The amount of these materials is in the range of 100 g/m of cable core. Replacement materials for carbon-black filled polymers are not known.

Most submarine power cables, especially those for rated voltages of 52 kV or above, are equipped with a metallic sheath. Lead alloy sheaths contain minute amounts of one ore more of the alloying elements antimony, tin, copper, calcium, cadmium, etc. Due to the large amounts of lead it constitutes a high value in the category "raw material depletion" in the environmental assessment. When enclosed in a plastic over-sheath the lead sheath will not release material into the ambient. Watertight sheaths made from aluminium or copper consume less material than lead sheaths, so they achieve better RMD scores. The environmental impact in different categories can be visualised in polar coordinate diagrams such as Fig. 10.1. The impact categories are distributed around the circle and the impact in each of these categories is shown by the distance to the centre. Figure 10.1 shows such a "radar plot" of two different land cable types [1]. The diagram is normalised, so that the values for the 1200 mm$^2$ Cu cable with Pb sheath are set to "1". Therefore,

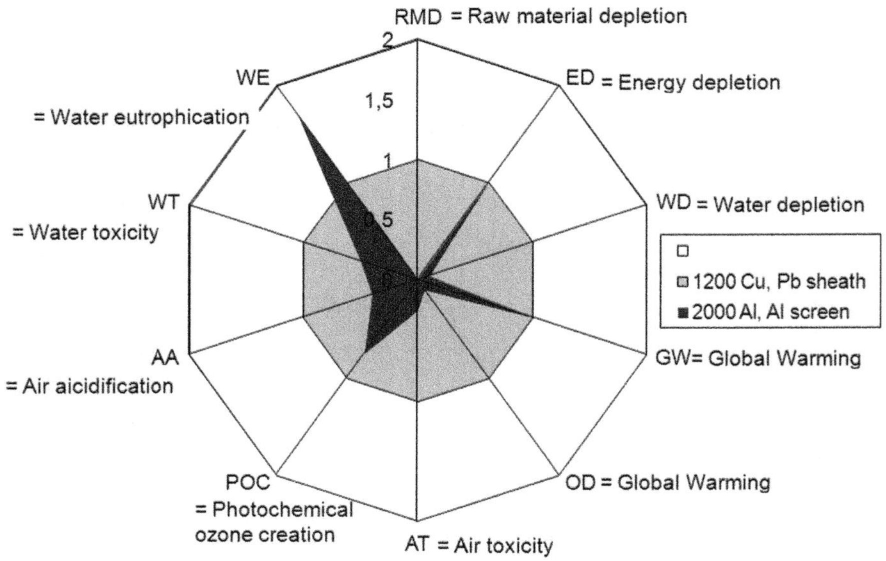


**Fig. 10.1** Radar plot over the environmental impact in various categories of two cable types

the profile curve for the Cu/Pb cable is at “1” around the circle. In most categories the Al/Al cable has much lower environmental impact than the Cu/Pb cable. The two cable types are on a par in the categories Energy Depletion and Global Warming, and almost on a par in Photochemical Ozone Creation. Only in the category Water Eutrophication the Cu/Pb cable is more environmentally friendly than the Al/Al cable.

The radar plot of Fig. 10.1 has been established for underground cables and may illustrate the assessment method. Similar plots, comparing submarine high-voltage a.c. and HVDC cables, however, could not be found in literature.

The armoring is most often made from wires of steel or non-ferrous metals. Design and cost considerations usually dictate the choice of material. All metallic armoring can in principle be recycled. Anticorrosive materials, such as bitumen, or plastic coatings, can render the recycling difficult. Some anticorrosive agents contain large amounts of solvents that contribute to the total emission of volatile organic matter during manufacturing.

The interstices of multi-core submarine power cables contain filler materials. These materials are most often polymeric ropes or extruded profiles made from PVC or PE. These materials are functionally almost equivalent so that the cable designer can make a choice from cost and ecological criteria.

Table 10.3 lists the content of some materials in two typical submarine power cables. In the HVDC case, the listed material amounts refer to a pair of single-core cables. The pair of HVDC can transmit more than twice the power compared to the a.c. cable. It becomes clear that HVDC cables can transmit much more power per invested material than three-phase ac cables.

**Table 10.3** Mass content of various materials in two types of submarine power cables. All mass figures in kgs

| | 170 kV a.c. 3 × 400 mm$^2$ Cu | HVDV 150 kV 1 × 1200 mm$^2$ Cu |
|---|---|---|
| Copper | 11 | 22 |
| Lead | 22 | 13 |
| Steel | 20 | 16 |
| XLPE | 8.3 | 3.8 |
| Other polymers | 5.9 | 2.2 |
| Tapes | 0.8 | 0.8 |
| Optical cables | 0.2 | – |
| Anticorrosion agents | <0.5 | <0.5 |

## 10.4 Environmental Aspects of Cable Installation

Like any construction activity, the installation of submarine cables affects the environment. Encouraged by the requirements of the authorities and the public opinion, the client will follow rules and standards in order to minimize the environmental impact.

All cable installation starts with the transport of the cable. As most submarine cable factories have an own port, cargo vessels offer a cost-efficient transport method even for short drum lengths. Cargo ships are known to have the best fuel economy per payload unit of all transport methods. Transport by truck has a larger environmental impact than sea transports.

It is taken as a prerequisite that the contracted transport and installation companies comply with industrial environmental standards also for their offshore activities. It is recommended to subject contractors to an HSE (Health, Safety, and Environment) audit. An established HSE incident reporting system enables a close follow-up.

The installation of submarine cables may comprise a few separated activities along the cable route. The first thing to do, after the desk-top studies, is a marine survey. In most cases, this is performed using minor survey vessels that do not impose a greater environmental impact than any other equally sized vessel. There are no reported indications that the used high-frequency sonar methods with low audible amplitude affect cetaceans or other marine mammals.

After survey, the cable route sometimes needs to be cleared of obstacles. Different methods can be used:

Pre-Lay Grapnel Run (PLGR): A hooking device (grapnel) is pulled along the seafloor prior to the installation operation. The hooking device is much smaller than most commercial fishing gear. If existent, seafloor vegetation can be uprooted in a very narrow corridor defined by the grapnel size. The PLGR is considered non-invasive to the seafloor.

Singular actions can be taken where abandoned cables or lines are cut; wrecks or other man-made obstacles are removed. Abandoned anchors and fishing gear *en route* are removed to tidy up the seafloor.

Rectification of the cable route. When the seafloor is littered with outcrops, boulders or other obstacles it can be difficult to find a manageable cable route between these obstacles avoiding free spans and other hazards. A close-up optical survey can be done by non-disturbing ROV operation. In earlier times the cable route was rectified by blasting a clear "road" through the debris on the seafloor. This method would not be accepted today in most waters.

The cable laying per se is a quite slow and calm activity. The CLV is travelling slowly along the cable corridor (at 0–3 knots) emitting the typical noise of vessels of similar size. Cable laying equipment such as capstan wheels, linear cable engines, etc. are installed on-board and do not contribute to noise outside the hull. No chemicals or gases are released during the laying except for the normal exhaust fumes from the vessel engines. The waste from normal vessel operation is stored and processed according to vessel standard. If the CLV is travelling along close-to-beach routes, where large vessels normally not operate, the beach-side dwellers might sense some low-frequency engine noise and exhaust smell during the duration of operation. The CLV also can disturb marine fauna especially during reproduction periods. Usually the permitting authorities will restrict the laying time frame avoiding periods of animal mating and breeding.

The cable is fed out from the CLV in a slow pace and moves down to the seafloor while the CLV is moving forward. Often the laying is monitored by ROV. ROVs use electric thrusters and bright lights that may frighten off animals and may disturb their behaviour for a short time. Some fish however seem to be very curious and look straight into the cameras of the ROV. Other animals can be observed while continuing with feeding under the close-up eyes of the ROV.

During the laying or afterwards ("post-lay trenching") most submarine power cables are buried into the seafloor. The burial by plough or waterjetting involves a vehicle moving on the seafloor either pulled by a surface vessel or by own propulsion. The trenching gear usually stirs up a lot of sediment in a narrow corridor (0.5–2 m) along the cable. Although most of the sediment settles back within short time, the sediment plume can, depending on sediment character and water currents, move hundreds of meters from the origin. Also, a local disorder in the natural layers can occur in the trenching zone. The degree of impact is depending on the trenching method and the soil conditions. The seafloor vegetation, if existent, will be disrupted in parts of this corridor. Benthos organisms[1] will be disturbed during laying operation, either by mechanical impact, or by scaring away. Animals living on these benthic organisms may leave the place with them.

Soft seafloors such as in the North Sea, or Baltic Sea, start to reorganize rather quickly after the burial operation. The trench refills itself unless it was cut in rocky soil. Mobile organisms like zoo benthos return to the area above the cable, and the vegetation recovers after a few growing seasons. In most cases the cable route is indiscernible within a year after burial. In the case of the SwePol HVDC cable in the Baltic Sea the zoo-benthos had returned completely within one year

[1]"Benthos" is the collective term for organisms living close to and in the seafloor.

after trenching [5]. The disturbances from the cable laying and burial are considered as temporary and small-size. Cable laying and burial is a singular event and should be evaluated in relation to other repeated disturbances by fishing gear ripping up the seafloor regularly. In the seafloors of the North Sea or the Baltic Sea that are being ruptured by trawling repeatedly, a single cable trenching would not make a difference. Comprehensive analysis and evaluation of the disturbances from cable laying can be found in the permits issued by the German marine authority BSH (Bundesamt für Seeschifffahrt und Hydrographie) for cable links to North Sea OWP [6].

However, there may be exceptions from this. Cable laying and burial in coral reefs and mangrove forests may cause irreversible damages. Every new cut can endanger the integrity of these biotopes. Other sensitive habitats such as sanctuaries for fish, mammals, amphibiae and waterfowl, might suffer damage, especially when the operation is being done during mating or nesting seasons.

Hazardous waste from hidden burdens, such as chemical dumps, ammunition dumps, also dumps of chemical warfare, can be stirred up and released into the sea water in critical areas during trenching. It is often not feasible to trench the cable in rocky seafloors. In that case, the cable can be protected by rock dumping or covering with concrete mattresses or something similar. The used materials must be compatible with salt water, neither must they leach out adverse substances.

Waterfowl react differently on the presence of the installation vessels. Some species are repelled for the time of installation, while others are attracted, as they would be by any other vessel. There is no obvious danger for waterfowl as long as the installation does not infringe their nesting habitat.

Concerns have been expressed for the noise impact on marine animals by submarine construction works. The sensitivity to noise of marine mammals is extremely difficult to access. The mammals live in environments that are not silent even in the absence of human activity. Marine mammals represent an extreme example of not only habitat adaptations but also adaptations in ear structure and hearing capabilities. Audiograms are available for only a few out of 119 different species of marine mammals, and these audiograms were recorded from captive individuals [7]. Marine mammals as a group have functional hearing ranges of 10 Hz–200 kHz. They can be divided into infrasonic balaenids (probable functional ranges of 15 Hz–20 kHz; good sensitivity from 20 Hz to 2 kHz), sonic to high-frequency species (100 Hz–100 kHz; widely variable peak spectra), and ultrasonic dominant species (200 Hz–200 kHz general sensitivity; peak spectra 16–120 kHz) [8]. Mammals use their voice and hearing capability for communications in the lower frequency range and for echolocation in the ultra-high frequency range. Most pinnipeds, such as seals, have their best hearing sensitivities at 10 kHz or above. Many documented examples show that cetaceans react on anthropogenic sound. Sound sources were sonar pingers, seismic equipment, vessel engines, industrial sounds etc. Behavioural changes in the presence of sounds range from shorter diving intervals, decrease or cessation of calls, avoidance of the place or escape. However, there is no evidence that these behavioural changes are biologically significant. Massive whale stranding has been associated to the use military of high-energy mid-frequency sonar. A clear demonstration of the causal relation is still to be done. Very little is known about the hearing capabilities of fish, marine invertebrates or marine reptiles.

For the case of noise generation, submarine cable installation should not be confused with other offshore activities involving hammering down poles into the seafloor.

## 10.5 Environmental Impacts from the Operation of Submarine Power Cables

### *10.5.1 Thermal Impact*

Power cables transmitting electric power dissipate losses in form of heat. The amount of heat losses at full transmission power is 10–100 W per metre of cable, which is corresponding to the heat from a household light bulb every metre. Inevitably the heat is transferred into the surrounding ambient. Cables installed on the seafloor do not heat up their surrounding as the water washes away almost all heat. However, the cable surface of the freely installed cable might be some degrees warmer than the surrounding water. Buried cables warm up the surrounding soil in the long run. Figure 10.2 shows the isothermal lines of temperature rise around a typical HVDC cable installation under the seafloor. It is a pair of 100 mm dia HVDC cables buried 1.0 m under the seabed. Each cable has a heat loss of 30 W/m. For the sake of clarity the isothermal lines are drawn only up to 26 K. The cable surface temperature is approx. 32 K above the seabed temperature.

The temperature rise in the immediate vicinity of the cable may alter the conditions for deep-dwelling cold-adapted organisms. In most cable projects this limitation is of no great concern as submarine power cables are designed to eliminate excessive losses. From subsea inspections and recovery operations it is known that many submarine power cables are colonialized by molluscs, starfish and other species. Probably submarine power cables, similar to artificial reefs made by OWP foundations or sunken ships, constitute new habitats in otherwise featureless seafloors.

### *10.5.2 The 2 k Criterion*

The losses in buried submarine power cables generate a temperature rise in the surrounding soil. If the cable is fully loaded over months the cable surface can reach a temperature of 40–50°C. This is, however, an extremely rare situation. Most submarine power cables have an average load much lower than the maximum rated load. The temperature decreases with the distance to the cable. A few meters from the cable there is practically no sediment warm-up (cf. Fig. 10.2). The

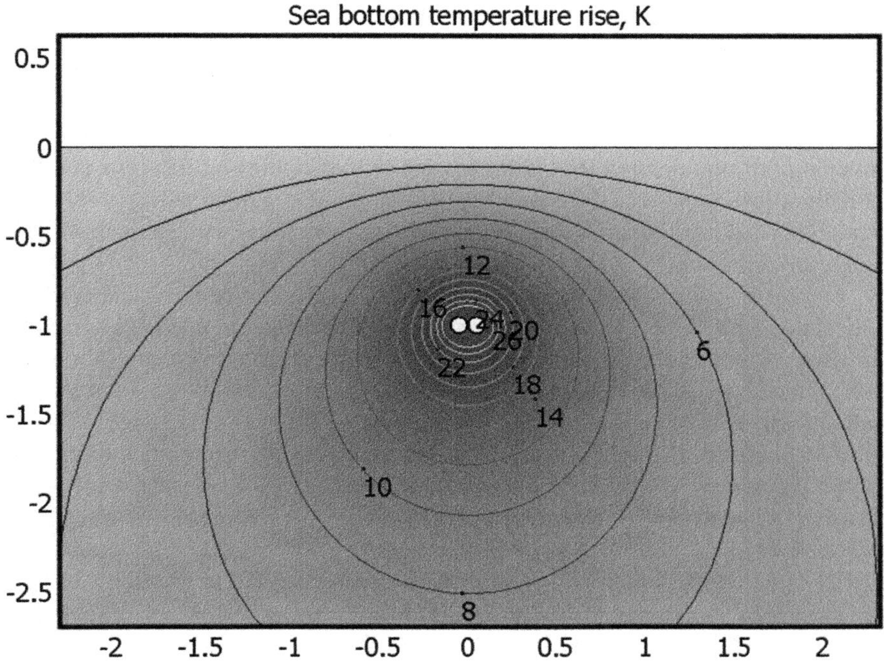


**Fig. 10.2** Isothermal lines around a pair of submarine HVDC cables. The horizontal line at "0" represents the seabed. The *y*-axis is the vertical position and the x-axis the horizontal position, expressed in m. The numbers in the diagram express the temperature rise above the temperature of the seabed, expressed in K. Spacing between isothermal lines is 2 K

changed temperature situation in the soil close to the cable alters the living conditions for *benthos.* Most of sub-seafloor dwellers live in the upper few centimetres of the seafloor. There are concerns that the temperature change could change the metabolism and reproductive behaviour of the zoo-benthos. Psychrophilic (cold-loving) organisms within the benthos might be expelled from the cable corridor, while new, foreign species may be attracted to the new "warm" habitat. Neither of this is wanted. Also, the contents of nutrients and oxygen can possibly be changed by altered temperature profiles. Still, there is no scientific evidence for any of these concerns. If there is a measurable impact of the local heat generation on marine life, it is limited to a very narrow corridor along the cable.

The question was discussed in greater detail when a number of submarine cable projects were filed for permission in the Wadden Sea off the German North Sea coast. The Wadden Sea is a national park and biosphere reserve recognized by UNESCO. Naturally, environmental issues came into focus when cable project applications arrived at authorities. To cope with the situation and to prevent possible damages to the environment, a German environmental authority (Nationalparkverwaltung Niedersächsisches Wattenmeer) has issued a recommendation that the temperature rise over a buried submarine power cable should not exceed 2 K[2] at a depth of 0.3 m below the seafloor, compared to the undisturbed seafloor. Another German environmental authority (Bundesamt für Naturschutz) has recommended a 2 K limit for a depth of 0.2 m, valid for the German Exclusive Economic Zone outside the 12-sm national zone. The German marine authority (BSH), which is in charge for the permits for submarine power cables in the EEZ, also defines a 2 K limit for a depth of 20 cm under the sea bottom [9]. Empirical support for this recommendation is not available. Perhaps the defined limits will be changed when evidence condenses.

The temperature rise at the –0.2 and –0.3 m levels, respectively, can easily be calculated with FEM software. Another simplified calculation method is described in Chap. 3, which can be performed with a simple pocket calculator. No matter which method is used, a good knowledge of the thermal resistivity of the seafloor and the expected cable load scheme is necessary.

What happens if a cable project does not meet the 2 K criterion as formulated above? A strategy to reduce the warm-up at the –0.2 and –0.3 m levels, respectively, is a deeper burial depth. Deeper down the cable has a smaller impact on the reference levels, everything else unchanged. For a single cable with 50 W/m losses and a burial depth (BD) of –1.0 m the temperature increase at –0.2 m would be 3.23 K.[3] To meet the 2 K criterion at –0.2 m the BD must be increased to –1.6 m. However, this strategy also has negative impacts on the environment. As outlined above, all benthos will be disturbed in a small corridor during the burial operation. Burying the cable down to –1.6 m would affect more benthos than a –1.0 m burial, both due to the larger depth and due to the wider corridor associated with a greater burial depth. Also, since a deep burial operation is more complicated than a shallow, the duration of the construction work is prolonged. The same applies for the recovery of the submarine cable after end of life.

Furthermore, the cable operation temperature (both on the cable surface and the cable conductor) increases. In the example above, the cable temperature is expected to increase with 4 K. At higher temperatures, the cable losses increase and destroy valuable electric energy. Using a greater burial depth to meet the 2 K criterion would result in extra environmental impacts, such as more seafloor disturbances, higher cable temperature, larger energy losses, and a larger volume of seafloor being subjected to cable heating. These effects should be accounted for when discussing the 2 K criterion in environmental assessments.

Another strategy to meet the 2 K criterion is the reduction of cable losses. For the situation described in the previous paragraph the cable at –1.0 m should have losses of 31 W/m rather than 50 W/m to meet the 2 K criterion. Loss reduction in submarine power cables can be achieved by different design measures. The most prominent is the investment in a larger conductor. In HVDC cables, the reduction from

---

[2] 1 K is one degree step on the Celsius scale.

[3] This has been calculated for a seafloor thermal resistivity of 1.0 K.m/W. To adopt the result to a real situation just multiply with the actual thermal resistivity.

50 to 31 W/m would require a conductor area growth by 61%. Increased conductor area would benefit the environment also by reduced energy loss, but is difficult to negotiate with investors. Furthermore, the environmental footprint of the production of the large conductor cable is obviously much larger than for the smaller cross section.

It is not the task of this book to argue the reasonability of the 2 K criterion. However, for the sake of better understanding, it could be useful to put the temperature rise of 2 K into perspective. The annual variation of the seafloor temperature can be as much as 15 K in shallow waters. It is also known that the seafloor temperature can vary considerably between different locations in the same area. That means that a seafloor which is 2 K warmer than another place in the neighbourhood, is probably not a "disturbed" environment. It seems that the environmental concerns would benefit from a wider view on the matter. Maybe a narrow corridor of 5 K warm-up (from a shallow cable) makes less harm than a wide corridor with 2 K warm-up and a larger heated soil volume (which would be the case with a deeper cable).

### *10.5.3 Electromagnetic Impact*

The nature of electromagnetic phenomena is not easy to grasp for non-electric engineers. The public is concerned about the exposure to all kinds of electromagnetic fields in daily life. Sometimes things are confused and pre-mature conclusions are drawn. Fortunately, the issue is not as complicated as it might seem. To start with, two properties of the electric and magnetic fields from power cables should be mentioned:

- Due to the low frequency, the fields around the cable do not constitute a "radiation" such as radiation from cell phones, antennas etc. The intensity decreases very strongly with distance.
- The fields from the cables contain so little energy that any ionising or cell disruptive effects are excluded.

For the consideration of power cables, electromagnetic fields can be separated into electric fields and magnetic fields. In the cable case, these fields are independent from each other except a few cases.

Electric fields are confined between conductive surfaces with a potential difference in-between. This is the case in power cables. The electric field exists between the high-voltage conductor and the grounded screen/armoring. As the screen/armoring is on the same electric potential as the ambient outside, there is no electric field outside the cable. The screen/armoring efficiently confines the electric field to the inside of the cable. This is valid for any power cable, regardless the design, rating, single-core or three-core cable, a.c. or d.c.

A secondary electric field effect is caused in the vicinity of a.c. cables. The a.c. current in the conductor creates a magnetic field outside the cable. This alternating $B$-field generates, according to the Maxwell laws, an alternating electric field with very tiny amplitude within the range of a few μV/m. This is much less than the electric field produced by the battery in the LED-lit key chain in our pocket. Some *elasmobranch* fish species such as sharks, rays, and skates, can sense these extremely small electric fields. They are assumed to use electric field sensation for orientation in the electric fields generated by water movement in the geomagnetic field, and for detection of prey. The *elasmobranch* detection level is within the range of 0.5 ... 1000 μV/m, the avoidance limit being assumed at >100 μV/m.

The magnetic field is somewhat different. Any conductor is surrounded by a magnetic field generated by the current. The intensity of the magnetic field at a certain distance from the linear conductor is according to the Biot-Savart law:

$$B = \frac{\mu_0 \cdot I}{2\pi \cdot d}$$

where $B$ is the magnetic field[4] at distance $d$ from the conductor, $I$ is the conductor current,[5] $\mu_0$ is the magnetic constant $\mu_0 = 4\pi\ 10^{-7}$ T·m/A, where T (Tesla) is the unit for the magnetic field (also called magnetic flux density).[6] As an example, the magnetic field at 1 m distance to a conductor carrying 1000 A would be $2{\cdot}10^{-4}$ T = 200 μT. Most often HVDC cables are installed in pairs with currents flowing in opposite direction. Therefore their magnetic fields are also anti-directional and cancel each other to a large extent, depending on the distance between the conductors. As a result, the magnetic field decreases very rapidly as a function of distance from the cable pair. Figure 10.3a shows the magnitude of the magnetic field vector over a pair of HVDC cables carrying 1000 A each. The cable pair is installed at the level –1.0 m under the seafloor. The axial distance between the two cables is 0.1 m, i.e. they are laid touching each other, or close to. The lateral distance from the centre axis between the cables is drawn on the $x$-axis. The upper curve illustrates the B-field on the seafloor. The next curve is the B-filed at the +5 m level and the lowest curve represents the B-field 10 m above the seafloor. Here the B-filed is far below 1 μT. Figure 10.3b shows the magnitude of the magnetic field vector over a pair of HVDC cables laid with 10 m distance to each other. Again, both cables carry 1000 A in opposite direction. The sequence of the three curves is equivalent to that in Fig. 10.3a. At the seafloor, 1 m above the cables, the B-field reaches values of 200 μT. At 10 m above the cable pair the B-field has decreased to 10 μT.

Most countries have developed recommendations or limitations for human exposure to electromagnetic fields. In 1992, the International Commission on Non-Ionizing Radiation Protection (ICNIRP) was established to investigate possible

---

[4]Strictly spoken, the symbol $B$ stands for the magnetic induction with the SI unit Tesla, while the magnetic field is denoted $H$ with $B = \mu_0 H$. However, in power engineering context $B$ is called magnetic field.

[5]The Biot-Savart formula assumes that the current flow is confined into a filament with zero diameter. For some distance from the cable conductor (>10 × conductor diameter) the Biot-Savart law can be used without any practical restrictions.

[6]The "Tesla" is a unit of the magnetic induction. Another unit is "Gauss". 1 T = 10,000 G = 1,000,000 μT.

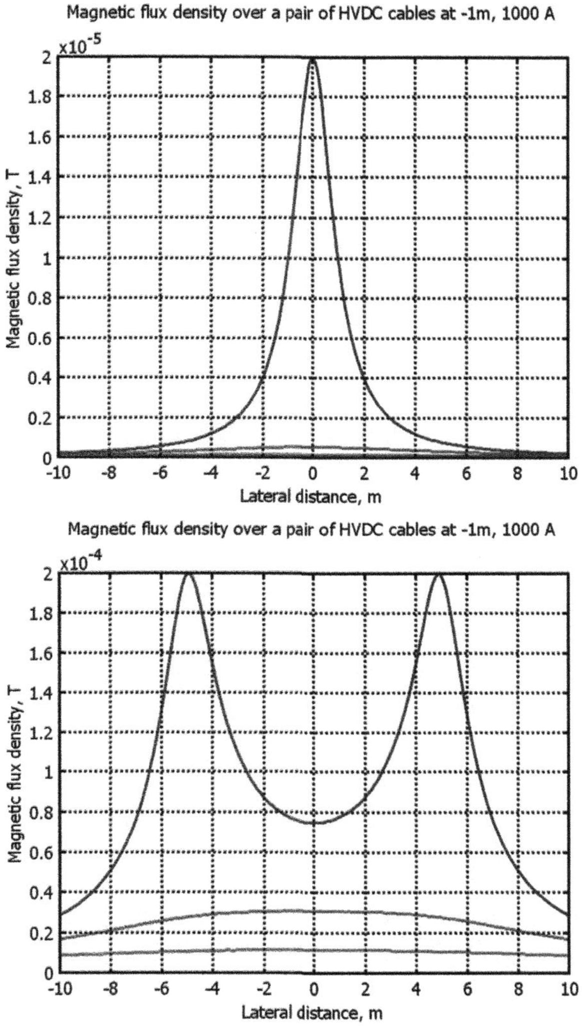


**Fig. 10.3** Magnetic field around a pair of submarine HVDC cables carrying 1000 A. As the magnetic field is proportional to the current, the results can easily be scaled up or down for other current values. Upper diagram: cable pair installed touching. Lower diagram: cable pair installed with 10 m distance

hazards from electromagnetic fields, and develop international guidelines in radiation exposure. This work resulted in the "Guidelines for limiting exposure to time-varying ... electromagnetic fields..." [10]. The guidelines cover frequencies up to 300 GHz (radio signals, mobile phones etc.), but for submarine power cables only two frequency ranges are relevant: 0 Hz (d.c.) and 50/60 Hz (a.c.). The recommended limits for human exposure for these frequencies are given in Table 10.4. Most European countries and Singapore, South Africa, South Korea, and Taiwan have adopted the ICNIRP values.

It is obvious from the diagrams that the magnetic fields emitted from d.c. cables are far below all known limits, even in close vicinity. Therefore no risks for human life or health can be anticipated from the magnetic field of HVDC cables. Concerns

**Table 10.4** ICNIRP Guidelines for magnetic fields

| | Occupational exposure | Public exposure |
|---|---|---|
| 0 Hz (d.c. cables) | 200,000 μT | 40,000 μT |
| 50 Hz (a.c. cables) | 500 μT | 100 μT |
| 60 Hz (a.c. cables) | 417 μT | 83 μT |

have been raised on the impact of d.c. magnetic fields from HVDC cables on submarine organisms, in particular fish relying on geomagnetic orientation during reproductional migration. Investigations carried out by the Swedish Fishery Authority demonstrated that some fish species obviously are not affected by the magnetic field of the SwePol cable, one of the highest rated submarine HVDC cables. Eel, known for its ability to navigate using the geomagnetic field, showed no abnormal behaviour when crossing HVDC cables.

Sometimes the magnetic field from HVDC cables is compared to the geomagnetic field, which is considered harmless to organisms. The intensity of the geomagnetic field is different at each location on earth but it is within the range of 0–50 μT. More serious than the impact on animals of magnetic fields from d.c. cables is their possible influence on the function of magnetic compasses. Magnetic compasses rely on the geomagnetic field and can be disturbed by other magnetic sources. Magnetic compasses can suffer a deviation in the immediate vicinity of powered submarine HVDC cables. The amount of deviation is depending on:

- Distance between the conductors of a cable pair
- Magnitude of the d.c. current
- Vertical distance between cable pair and compass localisation. The distance is the sum of burial depth, water depth, and height of the ship's bridge over the water surface. On large vessels, the bridge can be some 10–30 m over the water surface
- Magnitude and orientation of the local geomagnetic field
- Cable route heading. An East–West cable produces magnetic fields in N–S direction, which add to the geomagnetic field but do not alter the direction. The impact on magnetic compasses is much less than from a N–S cable under similar conditions.

For a small vertical distance between cable pair and the compass, considerable deviations (30° and more) can be observed directly over the cable pair. The corridor of noticeable disturbance along the cable is only a few tens of meters m wide.

It could be demonstrated that ships steered by autopilots based on magnetic compasses may suffer considerable route changes when passing over a single-pole HVDC cable [11]. When the vessel travels towards an HVDC cable the increasing deviation may trigger the magnetic autopilot to take action to correct for the perceived erroneous heading. In the worst case, the autopilot will steer the vessel so that it follows the cable route; the vessel is "captured" by the cable. Tests with a 17 m vessel with non-metallic hull were conducted over the Kontiskan cable in monopolar operation at 1200 A and 20 m depth [11], and confirmed the risk of "capturing". This risk is larger if the vessel heading at small angle to the cable route before crossing. In less onerous cases the autopilot steers the vessel over the cable route with unchanged heading but with a lateral displacement. Vessels are affected in different ways depending on their size.

Small vessels such as leisure vessels navigate shallow waters where the vertical distance between cables and compass can be very small. The magnetic compass of these vessels may suffer large deviations in the immediate vicinity of the HVDC cables. Normally, these vessels are navigated also by visual orientation and/or GPS based systems, which are not affected by magnetic disturbance. In a risk evaluation, the additional risk due to compass deviation seems small compared to risks from ignorance and bad seamanship. The consequences of a vessel accident are quite small.

Medium size vessels (draught 2–4 m) can normally not navigate into the shallowest waters. The magnetic compasses of these vessels can still be affected by magnetic deviations. The result can be a lateral set-off of the vessel's position. This would be serious in a ship separation lane, where the vessel, involuntarily, could be steered into the lane of opposite direction. But in ship separation lanes the water depth is large enough to provide a sufficient vertical distance between the HVDC cables and the magnetic compass.

Large vessels (draught $< 5$ m) have usually a large vertical distance between the compass position and the cable pair. These vessels have a long response time for course changes. They travel through the influence corridor before the magnetic autopilot can turn the vessel very much. The risk for these vessels is considered very low.

Only magnetic compasses and magnetically controlled autopilots can be affected by cable's magnetic field. Almost all commercial vessels have redundant gyrocompass or GPS based navigation and steering systems, which are not affected by magnetic fields.[7] HVDC cables do not impose a risk to marine traffic. Still, some countries require that commercial vessels have magnetic compasses as back-up systems.

In a.c. cables the current in each conductor is changing its direction 50 or 60 times a second, and so is the magnetic field. In three-phase cables the three magnetic fields from the conductors are superimposed to each other. For an observer at large distance (called "far-field", e.g. 25 times the cable diameter) all three conductors seem to have the same distance and direction. Hence the three magnetic fields cancel out each other almost perfectly. To a near-field observer close to the cable, the conductors are not exactly at equal distance and hence their magnetic field does not cancel out each other's perfectly. A minute residual magnetic a.c. field can be measured.

---

[7] Some GPS use incorporated magnetic compasses to detect global directions when the vessel is not moving and the NSWE direction cannot be determined from incremental positioning.

Single core cables often have a much larger distance between the conductors resulting in a poor cancelling of magnetic field at least in the near-field range. However, in single-core a.c.cables the conductor current induces a counter-directional current in the screen of the cable, resulting in a largely decreased external magnetic field.

### *10.5.4 Chemical Impact*

Under normal operation conditions submarine power cables do not release chemicals, consumables, or other agents to the ambient. The materials are designed to be stable under the influence of seawater for decades. Cables with solid insulation (XLPE, EPR, mass-impregnated) do not contain fluids that can leak in case of cable damage or rupture. The impregnation compound in mass-impregnated HVDC cables is so viscous (thick-flowing) that there is practically no outflow from the cable even if it is cut [12].

A different situation arises for oil-filled cables with low-viscosity insulation oil. These cables are now installed only for the highest voltages (230 kV and above) as they have a long operational record. The oil inside these cables is pressurized from onshore feeder stations, and pours out into the sea when the cable is damaged. In case of damage the cable owner tries to counteract the intrusion of water by pumping in more oil into the cable. It can take days and weeks before the cable leakage can be localised, and the cable can be picked up and sealed. A damage on a 420 kV Öresund cable between Sweden and Denmark in 1979 could be localized only after observations of the oil plume [13]. When an oil-filled cable on the Oslo fjord was damaged in April 2008, oil leaked out from the cable initially at a rate of 80 l/h. When one of the 138 kV Long Island cables was damaged by a ship anchor, the initial outpour of cable oil was 450 *l* a day; it took 10 days to seal the faulty cable [14]. The good thing is that the cable insulation oils have so low viscosity that they usually evaporate rather quickly from the water surface. The cable insulation oils used to be low viscosity mineral oils. Since the 1980s synthetic cable oils have been used both for refurbishment and production of cables. Most of the synthetic oils are linear alkylbenzenes (LAB) with varying molecular weight composition. LAB is miscible and compatible with mineral oils. LAB is biodegradable, but little is known on the rate of degradation under the exclusion of air. The toxicity of LAB has been demonstrated by tests showing that the $LC_{50}$ mortality concentration in soil for a nematode species is less than 1% [15]. A study on the subject and further references have been published in [16].

## 10.6 Recycling of Submarine Power Cables

There are few options to proceed after the useful life of a submarine power cable. It is easy to leave the defunct submarine cable in place and announce it as "abandoned" to authorities. It is difficult to pull up buried cables, and the scrap value might not pay the salvage costs. Abandoned cables continue to be an obstacle for fishermen (a late revenge of the cable owner perhaps?) and must be cleared before new cables are being laid. If it is decided to leave the cable in place, it is possible to remove the insulation oil by subsequent flushing the cable with solvents and fill it finally with water [17].

Another solution, sometimes also required by permit authorities, is the recovery of the cable from the sea floor. Recovery of buried cables can be difficult. Even heavy-armored cables can break under the load of a pull-up operation, but recovery is possible. As an example, 76 km of 138 kV submarine power cable were removed from the Georgia Straight between Vancouver Island and the Canadian mainland in 2007 after 51 years of service [18]. Once the cable has been recovered from the seafloor, it can be recycled in a number of different ways. The first step of every recycling is the separation of materials. The metal fraction in the cable (steel, copper, lead, and other non-ferrous metals) can be separated easily by mechanical disintegration and separation using the large difference in specific gravity of metals versus other materials. The value of the metals can partly pay for the cost of the cable recovery. A promising method to cut open long lengths of cable is the use of water-jets running along the cable.

The clue to a successful recycling of polymeric components (polyethylene, XLPE, polypropylene, PVC, EPR, EPDM, etc.) is the separation of the cable constituents into clean material streams. A sink-and-swim separation in water creates a polyolefin stream with PE, XLPE, PP, and the "PVC" stream containing conductive materials, PVC, EPR, polyester, etc. A material recovery (using it for production of new materials) is only possible with very clean homogeneous streams. The chemical recycling of silane-linked XLPE has been demonstrated in Japan [19]. The most cost-efficient and environmentally wise methods seem to be the use of the cable polymers as fuel in thermal power plants normally fuelled with waste or biomass. A comprehensive overview on recycling possibilities can be found in [20].

No methods have yet been reported to re-use paper insulation. Paper insulation makes up a good fuel for biomass fuelled power plants or district heating plants.

The third solution is the use of submarine cables to build habitats for marine life, or artificial reefs. It has been observed many times that organisms, such as molluscs, algae, starfish, corals, crustaceans, etc., settle or live on submarine cables. Marine organisms are in general attracted to items different from a flat seafloor, so artificial reefs have been built from all sort of structures. Of course, the materials must be cleaned from all residual fluids and other hazardous items. Both defunct cables and defunct cable laying ships have been used for this purpose [21].

## References

1. Mirebeau P et al. (2007). Use of an ecoconception software to design a hv cable connection, Proceeding of the 7th International Conferences on Insulated Power Cables, Jicable, Paris, France, 2007, 237 ff.
2. Drugge B et al. (1996). Life Cycle Assessment on High-Voltage Power Cables, Cigré session paper 21-206, Cigré Session 1996, Paris, France.

3. Report by the Swedish Power Research Institute Elforsk AB, http://www.elforsk.se/elforsk/miljovardering_elanvand.pdf
4. Nill M et al. (2001). Grundlagen zur Beurteilung der Nachhaltigkeit von Energiesystemen in Baden-Württemberg. Report submitted at a seminar at Forschungszentrum Karlsruhe, Germany. 28.2.–1.3.2001 http://bwplus.fzk.de/berichte/ZBer/2001/ZBerbwa99001.pdf
5. Andrulewicz E et al. (2003). The Environmental Effects of the Installation and Functioning of the Submarine SwePol Link HVDC Transmission Line: A Case Study of the Polish Marine Area of the Baltic Sea, Journal of Sea Research, No. 49, pp. 337–345.
6. Permission Issued by BSH for the Installation and Operation of a Submarine HVDC Link to Connect the OWP project "Nördlicher Grund", http://www.bsh.de/de/Meeresnutzung/Wirtschaft/Windparks/Netzanbindung%20Noerdlicher%20Grund.pdf
7. Committee on Potential Impacts of Ambient Noise in the Ocean on Marine Mammals: Ocean Noise and Marine Mammals, Ocean Studies Board, National Research Council Of The National Academies, ISBN-10: 0-309-08536-5, The National Academies Press, Washington, DC.
8. Wartzok D and Ketten D R (1999). Marine Mammal Sensory Systems.In Biology of Marine Mammals, J E Reynolds III and S Rommel (eds.). Smithsonian Institution Press, Washington, DC, pp. 117–175.
9. Federal Maritime and Hydrographic Agency (BSH): Standard. Design of Offshore Wind Turbines, BSH Doc no 7005, Hamburg/Rostock 2007.
10. ICNIRP (1998). Guidelines for Limiting Exposure to Time-Varying Electric, Magnetic, and Electromagnetic Fields (up to 300 GHz), Health Physics, Vol. 74, No. 4, pp. 494–522.
11. Akke M et al. (1995). Nonlinear Interaction of dc Cables and Magnetically Controlled Ship-Steering Autopilots, IEEE Transactions on Control Systems Technology, Vol. 3, No. 4, December 1995.
12. Worzyk T (1996). Keine Gefahr bei Lecks in Gleichstrom-Seekabeln, Elektrizitätswirtschaft, Vol. 26/96.
13. Elgh L et al. (1982). Repair of 400 kV ac Submarine Cable, Paper 21-05, Cigré Session 1982.
14. Pirelli Cable Review, June 1971.
15. Johnson S J et al. (2004). Toxicity Testing of Linear Alkylbenzene Cable Oil Using Toxicity Testing of Linear Alkylbenzene Cable Oil Using Caenorhabditis Elegans Caenorhabditis Elegans In Soil, Submission A13-30 Society for Experimental Biology Meeting 29 March–2 April 2004, Heriot-Watt University, Edinburgh, Scotland.
16. Johnson S J et al. (2001). Potential for Anaerobic Biodegradation of Linear Alkylbenzene Cable Oils: Literature Review and Preliminary Investigation, Land Contamination & Reclamation, Vol. 9, No. 3, p. 279ff.
17. Le Poidevin G et al. (2007). Advances in the Decommissioning of End-Of-Life Oil Filled Cables, Paper 0357, 19th International Conference on Electricity Distribution, May 2007, Vienna, Austria.
18. Suen H (2007). Removal of 76 km 138 kV Gas-Filled Submarine Cables with 50 Years Service, Presentation at IEEE PES Insulated Conductor Committee Meeting, Scottsdale, Arizona, USA.
19. Tokuda S et al. (2003). Thermoplasticizing Technology for the Recycling of Crosslinked Polyethylene, Furukawa Review, No. 23, pp. 88–93.
20. Hagström B et al. (2006). Disposal of Cables at the End of Life, IEEE Electrical Insulation Magazine, Vol. 22, No. 2, pp. 21–30. March/April 2006.
21. Drabble R (2006). Submarine Cable Decommissioning: Assessing the Environmental Risks. Article published on TCMnet.com: http://www.tmcnet.com/news/2006/03/03/1427163.htm

# Chapter 11
# Anecdotes

## Contents



The professional life of the cable engineer seems to be boring, maybe even desolate. The product he (in rare occasions: she) is working with has reached great technical maturity with no high-tech appeal whatsoever. Submarine power cables have no aesthetic value as no manufacturer so far has spent money to print small fish or sweet little anchors on the outer sheath. The product is invisible in the public life – like a sewage pipe. May be even lesser known to the great public. In contrast to the rocket scientist and the TV talk show leader the cable engineer would speak rather quiet about his/her profession in the dancing hall or in the flirt line.

T. Worzyk, *Submarine Power Cables*, Power Systems,
DOI 10.1007/978-3-642-01270-9_11, © Springer-Verlag Berlin Heidelberg 2009

Still, sometimes laughter can be heard during coffee breaks in the cable factories, the electro-technical institutes, or during expert conferences. The reason might be a funny story from the strange world of submarine cables. Some anecdotes and little tales are printed in this chapter. All this material is compiled from rumors and tales, sometimes from printed matter[1]. The author hopes to receive more and better stories from the vast treasure out there, for the next edition of this book.

## 11.1 The Floating Hospital S/S Castalia

The steamship S/S Castalia was built in order to facilitate the Dover-Calais traffic in the 1880s. After a few trial tours it was decided to de-mobilise her and commit her to other duties. Laid off, she was intended to serve as a pest hospital vessel near Dartford, England.

For matters of convenience the vessel should be equipped with electrical illumination. Four submarine cables with extra heavy armoring and an insulation made from vulcanised bitumen were produced by the "Callender Bitumen Telegraph and Waterproof Company". The year is 1889. Vulcanised bitumen had been patented in 1881 and had been used successfully in land cables.

The cable laying was successful onshore. But when the workers got to know that the cable was to be laid to a pest carrying ship, they terminated the work and preferred to attend the near-by "Long Reach Tavern". The duty engineer, George Barnard, soon realised that all attempts to persuade the crew to continue working would be futile.

He rowed the pull wire to the pest hospital vessel, and let the hospital staff pull in the cables. It is untold if also the sick were told to pull. The engineer himself concluded the cable termination work onboard alone [1].

## 11.2 HVDC Cable Between Lydd, UK and Boulogne, F

*"C/S Dame Caroline Haslett" left Woolwich Cable Works at 24 May (probably 1961) for laying the English part of the cable. She made preps and test runs at the mooring position off shore. At the same time a shore gang made final preps to pull the cable inshore.*

*The weather which had been perfect for the French operation, turned worse. It delayed the whole thing and added to the difficulties.*

*On 7 June the ship was fixed on a mooring point 1400 yards offshore. The cables were pulled out with pull wires towards the beach (winch on beach). Cables supported by air cushions (rubberized canvas air cushions). Cable pull out at 40 ft/min. Suddenly (when 360 yards were out) one air cushion and then two more broke away from the cables.*

*Chain reaction: Under the increasing weight of the cable all the other cushions were dragged under water and the whole cable pair disappeared.* [2]

---

[1] Text in *Italics* is reproduced directly from the cited references.

## 11.3 The Pilot

I had overseen the design, testing, and part of production of this powerful submarine cable. Now I was on board one of the largest cable ships of the world with more than 6000 t capacity, to watch the installation. After years of woes and whoops in manufacturing and testing I would observe how "my" cable would go smoothly into the water.

I sat on the bridge and watched the "pilot". The pilot is the cable laying chief. He/she commands the speed of all cable laying machinery, the vessel speed, and all other things relevant for the cable laying. By radio he/she controls all the slave cable engines and has the full responsibility for the cable laying operation. As always, the night shift included a certain amount of tiredness and sleepiness. To keep blood pressure positive people on the bridge usually start telling stories and fairy-tales. So I started to talk about an anecdote connected to the Leyte-Cebu cable system, which has been delivered by my company and another one. The Leyte-Cebu system included four single-core cables. Rated 230 kV, the cables were big guys. They were laid in parallel corridors about 1 km apart to make service possible if one cable should brake.

Shortly after laying, all four cables were reported broken and possibly disrupted. Disbelieve and incredulity were the main tempers in the offices of the cable factories. To prevent water from ingress, oil was pumped into the cable from the shore stations at a rate making necessary the air-borne import of new cable oil. It was a terrible mess altogether. During coffee breaks the possible cause of the breaks was discussed, ranging all between dynamite fishing and foreign sabotage. My eight-year-old son suggested a submarine volcano outburst as an explanation.

Later on the real cause became known. Another cable ship, laying a telephone cable for the Philippine Long Distance Telephone Co. across the power cable route had actually fresh sea charts but no information about the newly installed 230-kV cables. The tension curve from the laying displayed four neat peaks with one kilometre or so distance while the telephone cable was ploughed through the power cables.

The pilot listened to my narration with little interest while he watched the data monitors. "My" cable continued to roll out smoothly, and I felt in safe hands. He said, "And do you know who was the pilot on that telephone ship? It was me!"

## 11.4 S-Lay and Coiling Direction

Each trade has its own unwritten laws. Young engineers entering a design department of a radio manufacturer learn very early that volume knobs on the stereo always are to be turned clockwise to increase the volume. Why is that so? Is there a technical raison-d'être? The junior often does not dare to ask, anxious to avoid disparaging answers.

But why is it as it is? Why do we usually put an S-lay armouring on the cable so that it must be coiled clockwise? Can we do the other way round? Many things come in a left-hand and a right-hand appearance, e.g. screw threads, electric guitars (after Jimi Hendrix), or road lane use. Left-hand driving empires and right-hand driving empires have risen and fallen no matter what.

The S-lay (or left-handed lay) armoring has been described as standard already in 1898 [11.3]. The notable Messrs. R. S. Newall & Co. and Messrs. Glass & Elliot, the first to apply iron wire armouring to electric cables, were in the wire rope business, where right-handed lay was tradition. No question, they produced the first armoured telegraph cables with right-handed lay. However, this lay direction soon led to some unexpected inconveniences. The coiling method of the day was to start coiling from the outer rim of the tank towards the inner hub. Once arrived there the cable would be laid straight outward again, only to start the next flake also from the outside. The guys in the coiling tank, the coilers, now were told to lay the cable clockwise, to accomplish for the armouring lay direction. The coilers found this very inconvenient as they were forced to guide the cable with their *left* hand while they were running around the coiling tank. The only way to overcome this inconvenience was to change the lay direction of the armouring.

Some cable makers listened to their workers later than others. The first transatlantic telegraph cable was made partly by Newall (right-handed lay) and partly by Glass & Elliot (left-handed lay). A splice between these cables would result in undesired untwisting under the influence of strain. This was a major engineering glitch, today possibly called Insufficient Interface Coordination. We do not know what was said in the board room when the mismatch was discovered but I and my fellow

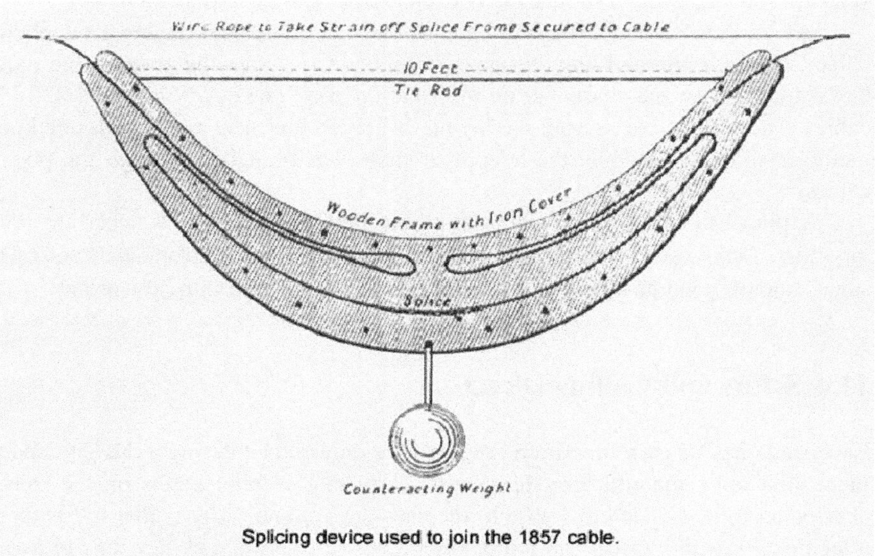


**Fig. 11.1** Transition joint between *left-hand* and *right-hand* armoured submarine telegraph cable of 1856 [4] (Courtesy of Bill Burns)

cable engineers only can bow our heads to the ingenious solution conjured up in these days (Fig. 11.1).

## 11.5 Edible Insulation

There were no given insulation materials in 1875 when the students of the Cornell University wanted to install electric illumination on the Campus. They were in search of a good yet affordable insulation material. It was the time when photography was brought forward by "kitchen chemistry" to find ever better development processes. Under the guidance of their electrical professors the students decided to build up the cable insulation from muslin, a fine fabric. The cable core was impregnated in beef drippings obtained from the local butcher [1]. Unfortunately, no records on the success of this strategy have been found yet.

## 11.6 Flipper

The North Sea is not always a cosy place for vacation. Many are the cable technicians who have fed the fishes during cable installation campaigns, especially during winter storms. The waves roll in relentlessly under the vessel, making even simple moves on board difficult. Not only easy tasks, such as balancing egg and bacon to the breakfast table, require full concentration but also cable work on deck or in the jointing house can be close to impossible when the vessel is rolling and swaying along.

J. S. and his team had finished a beautiful jointing job on the North Sea. The vessel was about to lay the next length of the power cable. It was decided to transfer the jointing team to shore for other duties. The vessel had no helicopter platform, so a transfer vessel was called in for taking up the jointers team. It is always a tricky thing to enter a rather small transfer vessel from a large vessel. The transfer vessel would approach the large vessel from the lee side, positioning its bow as close as possible to the large vessel. The man to be transferred would hang on a rope ladder outboard of the large vessel, watching the heaves of the bow of the small transfer vessel. The trick is to jump over to the transfer vessel exactly when the bow stands still for a split second on the top of its movements.

J. S. prepared himself for the transfer. His barrel-shaped body filled his orange survival suit completely without as much as a cubic inch of air inside. Familiar with the procedure, J. S. watched carefully the heave of the transfer vessel. After a few wave movements he stepped over on the bow of the transfer vessel. And now something unexpected happened.

Unforeseeably, another wave arrived under the transfer vessel, lifting up the bow with large force and speed. The bow went up like a catapult, sending J. S. right up in the air like an orange Michelin Man. J. S. flew a beautiful ballistic curve before he, head first, smashed into the water. While J. S. was picked up very quickly by the transfer team, a Danish voice rated the flight: "Nine point three!"

## 11.7 Stamps

The production and laying of submarine cables, may it be telecom or power, is a century-old art mostly unknown to the public. Still, from time to time, the achievements are brought to general attention, and why not by stamps. Submarine cables of any kind are connectivity, and this can be an enormous step forward to island nations, or countries that can be festooned with submarine cables. However, it is not really clear why countries such as Rwanda and Bhutan (each far away from the coast line) memorize submarine cables on stamps issued by their post offices (Fig. 11.2).

**Fig. 11.2** The "Great Eastern" on a stamp from Bhutan, 1989. The "Great Eastern" laid the first transatlantic telegraph cable in 1858

## 11.8 Unusual Cable Ships

The "Great Eastern" shown in Fig. 11.2 was the largest ship in the world of those days, mobilized for this truly great cable-laying job. Not every submarine power cable project of more recent date did enjoy the participation of such an impressive vessel. Cable layers do not need the luxury of the biggest and brightest but make do with anything that can serve their purpose.

**Fig. 11.3** (**a**) Purpose-built cable laying arrangement for calm waters. (**b**) Floating cable drum used as laying vessel. (Courtesy of NKT Cables A/S)

Figure 11.3a looks like a cosy installation job in calm waters, a colourful sunshade inviting to an after-hour-drink. However, the howl of the hydraulic gear would make the conversation somewhat inconvenient. Still, this cable laying spread can do its job perfectly well.

The Danish cable manufacturer NKT has successfully used floating cable drums for cable installation. Figure 11.3b illustrates that cable laying vessels can indeed come in many shapes. The floating drum was used for the installation of the Oresund cables between Denmark and Sweden in the 1950s [5]. Obviously, this technique was appropriate for short cable lengths and installation under vacation-like bathing weather, though the Oresund cables were installed in winds up to 10–12 m/s.

Cable installers would probably love to be enrolled on a job using a blimp (airship, Zeppelin) for their task. It is an amazing imagination sitting aloft over hostile swamplands with gaping crocs and poisonous frogs, paying out cable in a pleasant pace and enjoying an inspiring sunset drink. This is what the inventors of the patent application shown in Fig. 11.4 must have dreamt of [6].

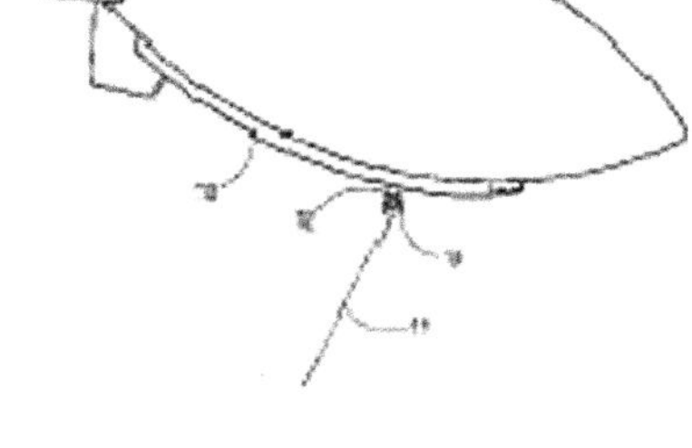

**Fig. 11.4** Cable laying airship illustrated in a patent application

## 11.9 Master Teredo

Not unlike a science fiction creature, the teredo worm is a very nasty submarine animal eating almost everything in its way. It has a rasp-edged shell, which it uses like a buzz saw to cut through wood – or through submarine cables. Cable companies learned the hard way, early on, that it likes to eat gutta-percha, an early plant-sap-based cable insulation.

The following text is taken from a book on submarine cable laying and repair published 1898:

> *"For protection against the teredo the Telegraph Construction and Maintenance Company serve the core with brass tape, a layer of white canvas tape ... This was introduced by Mr H. Clifford, the engineer of this Company, and has proved the most efficient protection ...*
>
> *For this purpose also, Bright and Clark's compound, containing mineral pitch, tar, and finally powdered silica is used in alternate layers with jute yarn outside the sheathing wires (sheathing means armouring in this text) by most manufacturing companies. From the success of this compound it would seem that the sensation felt by Master Teredo as his boring-fang touches the sharp glass-like grains is one which he does not care to experience again. The compound, in fact, was devised to render this part of his organisation useless for further exploits of the kind, at any rate on submarine cables."* [3].

Few, if any, cable engineers of today have the opportunity to give name to a particular cable constituent or production process. Still, some respected colleagues have lent their names to production methods. The Hjalmarsson degree of compactness of paper insulation is based on a sophisticated "kick-the-tires" with a cedar-wood pen. Paper insulation can be firm and compact (desirable) or weak and swampy (undesirable). The pitch of the sound when a wooden pencil is knocked onto the cable insulation indicates the quality to the experienced cable engineer. Old fellow Hjalmarsson taught us young guys this lecture.

## 11.10 Krauts at War Searching for a Cable Break

> *"My father worked in the cable industry in the 1950s and 1960s. He tells an interesting story from the start of the second world war (WW II).*
>
> *Immediately after Britain declared war in 1939, a cable ship was sent out into the North Sea to dredge up the German cables, which ran from Hamburg through the North Sea and out into the Atlantic. They found the cables using a hook, exactly as described in this story, and cut through them.*
>
> *Of course it would be easy for the Germans to go out with their cable ship, dredge up the two ends, and join them back together again – if they knew where to look for the break. And it's not hard to find out how far along the cable the cut is, as a pulse will be reflected from the break. This had been well understood for a hundred years.*
>
> *Knowing this, the British engineers made some sort of contraption full of capacitors and coils that they could fix on to the end of the severed cable before dropping it back into the sea. This would add some extra delay to the reflection, causing the Germans to miscalculate where the break was, and send their cable ship to the wrong place."*

Cited from Internet [7], also referred to in [8].

## 11.11 Even More Damages

When the first transatlantic telegraph cable showed poor performance due to weak signal transmission, two gentlemen argued about the right cure. One of them was William Thomson, later named Lord Kelvin. The other was Dr. Edward Orange Wildman Whitehouse, a medical doctor, who clearly worked in the wrong field. He suggested, and tried, to cure the cable with 2000 V shocks from his patented induction coils, such ruining the cable insulation for good.

> *Dr. Wildman Whitehouse and his 5-foot-long induction coils were the first hazard to destroy a submarine cable but hardly the last. It sometimes seems as though every force of nature, every flaw in the human character, and every biological organism on the planet is engaged in a competition to see which can sever the most cables. The Museum of Submarine Telegraphy in Porthcurno, England, has a display of wrecked cables bracketed to a slab of wood. Each is labeled with its cause of failure, some of which sound dramatic, some cryptic, some both: trawler maul, intermittent disconnection, strained core, teredo worms, crab's nest, fish bite, even "spliced by Italians."* [9]

## 11.12 Loops

When a large cable ship installed a large HVDC cable some years ago, the DP system was not really the latest and could hold only a limited number of pre-programmed way points. During the night, the officer on duty fell asleep for a short moment while the vessel automatically traveled along the cable corridor. All of a sudden, the last of the pre-programmed way points was reached and the vessel stopped. Unfortunately, pay-out of the cable continued and an unknown amount of cable was fed over the aft wheel until the vessel started move again. The cable laying operation was finished.

A later inspection however revealed a "beautiful" loop where the cable had been paid out without ship movement. Post-lay trenching was not possible at the location and the cable loop had to be protected by expensive shiploads of rocks.

## 11.13 Cable Ship Reefs

There are some submarine cable project managers who allegedly uttered a sigh combined with the strong wish to blow up the cable-laying vessel (CLV) involved with their project. For them, it may be a little comfort to know that CLV can be really useful as habitats for corals and fish once they have settled on the sea floor.

On July 25th, 2003, the Government of Caribbean St. Eustatius created its latest reef. The "Charles L. Brown", a 320 ft cable-layer acquired from AT&T, was sunk in 95 fsw and now lays on its side on a clear sandy bottom. The ship was originally constructed in 1954 in Italy and was acquired by Statia's government in 2002. The vessel was prepared for sinking (cleaning of oils and fuel, removal of hazardous waste, opening of safety passages for divers) in an operation managed by a committee of the three island dive operations, the Statia Marine Park and the Statia Tourism office. Today, a few short months after the sinking, divers can explore the wreck and observe as various marine organisms begin to make their home there [10].

The Navy cable ship USS Aeolus is one of the oldest artificial reefs. After a career as warship and cable vessel, she was sunk at her final resting place 22 miles south of Beaufort Inlet off Morehead City, North Carolina, in 1988. She sank on one side, but 1996 hurricane "Fran" turned the wreck through 90°, then righted here and broke here in three [11].

## 11.14 Poetry

In spite of their unimpressive appearance, submarine cables have been recognized in other literature than engineering books. The author of the Jungle Book, Rudyard Kipling, wrote a poem on submarine telegraph cables, transferring human messages through the unimaginably silent depth of the ocean. The poem goes as follows:

**The Deep-Sea Cables**

The wrecks dissolve above us; their dust drops down from afar –
Down to the dark, to the utter dark, where the blind white sea-snakes are.
There is no sound, no echo of sound, in the deserts of the deep,
Or the great gray level plains of ooze where the shell-burred cables creep.

Here in the womb of the world – here on the tie-ribs of earth
Words, and the words of men, flicker and flutter and beat –
Warning, sorrow and gain, salutation and mirth –
For a Power troubles the Still that has neither voice nor feet.

They have wakened the timeless Things; they have killed their father Time;
Joining hands in the gloom, a league from the last of the sun.
Hush! Men talk to-day o'er the waste of the ultimate slime,
And a new Word runs between: whispering, "Let us be one!"

## 11.14.1 The Journey of Mrs. Florence Kimball Russel

The figure shows the front cover of a book by Florence Kimball Russel, published in 1907. Mrs. Russel tells us about a long journey on a cable ship through the Philippine archipelago. Mission: installation of telegraph cables between the islands.

Mrs Russel writes:

*"Whether we are picking it up, or paying it out; whether it is lying inert, coil upon coil, in the tanks like some great gorged anaconda, or gliding along the propelling machinery into some other tank, or off into the sea at our bow or stern; whether the dynamometer shows its tension to be great or small; whether we are grappling for it, or underrunning it; whether it is a shore end to be landed, or a deep-sea splice to be made, the cable is sure to develop most alarming symptoms, and some learned doctor must constantly sit in the testing-room, his finger on the cable's pulse, taking its temperature from time to time as if it were a fractious child with a bad attack of measles, the eruption in this case being faults or breaks or leakages or kinks..."* [12].

No doubt, a book worth reading for a cable engineer. It not only pictures the worries of a cable installer, particularly at sea, but also reveals a great insight in the contemporary conditions in the archipelago.

Let me conclude this book with the words of Mrs Russel:

**Life on a cable-ship would be a lotus-eating dream were it not for the cable.**

## References


1. Hunter P V et al. (1956). "Development of Power Cables", George Newnes Ltd, London.
2. Black R M (1983). "The History of Electric Wires and Cables", Peter Peregrinus Ltd, London, ISBN 0863410014.
3. Wilkinson H D (1898). Submarine Cable Laying and Repairing, "The Electrician" Printing and Publishing Company, London.
4. Burns B, Private Communication, http://atlantic-cable.com/Article/WireRope/wirerope.htm
5. Geijer G et al. (1958). The Öresund Power Cable Project, Blue-White Series, Published by the Swedish State Power Board, 1958.
6. International Patent Application PCT/EP2003/003658.
7. http://slashdot.org/article.pl?sid=07/01/13/0750238&from=rss by "Anonymous Coward", January 13, 2007.
8. Singh S (2000). The Code Book, Anchor Books, New York. ISBN 0-385-49532-3.
9. Neil Stephenson "Mother Earth Mother Board", Wired, Issue 4.12 – December 1996, http://www.wired.com/wired/archive/4.12/ffglass_pr.html
10. http://www.ecophotoexplorers.com/StEustatius.asp
11. http://www.nc-wreckdiving.com/WRECKS/AEOLUS/AEOLUS.HTML
12. Florence Kimball R (1907). A Woman's Journey Through the Philippines, L. C. Page & Company, Boston. Reprints available on the Internet.

# Chapter 12
# Useful Tables

## Contents



## 12.1 Dielectric Properties of Cable Insulation Material

| | Dielectric loss factor tan δ | Dielectric constant $\varepsilon_r$ |
|---|---|---|
| Source: Anders [2] in Chap. 3 | | |
| IEC 60287 | | |
| XLPE ≤ 18/30 kV | 0.004 | 2.5 |
| XLPE > 18/30 kV | 0.001 | 2.5 |
| EPR | 0.005–0.020 | 3 |
| Oil/paper >87 kV | 0.0033 | 3.6 |
| Mass-impregnated | 0.01 | 4 |
| Source: Allister [8], p. 100 in Chap. 3 | | |
| XLPE | 0.008 | 2.5 |
| EPR | 0.04 | 3 |
| Oil/paper >87 kV | 0.004 | 3.3 |
| Mass-impregnated | 0.01 | 4 |
| Source: Bartnikas [9], p. 99 in Chap. 3 | | |
| XLPE 20°C | 0.002 | 2.3 |
| EPR | 0.0013–0.0023 | 2.7–2.8 |

T. Worzyk, *Submarine Power Cables*, Power Systems,
DOI 10.1007/978-3-642-01270-9_12, © Springer-Verlag Berlin Heidelberg 2009

(*continued*)

| | Dielectric loss factor tan δ | Dielectric constant $\varepsilon_r$ |
|---|---|---|
| Source: Deschamps [1] | | |
| LDPE | $8.10^{-4}$ | 2.3 |
| HDPE | $8.10^{-4}$ | 2.3 |
| Oil/paper | $30.10^{-4}$ | |

Note: Tan δ is depending on temperature.

## 12.2 Lead Alloys

| Alloy designation acc. to | | Alloy elements and percentage (by weight). Min and max values | | | | | | |
|---|---|---|---|---|---|---|---|---|
| EN50307 | Convention | As | Bi | Cd | Cu | Sb | Sn | Te |
| PK008S | PbSb-0.5 | | | | | 0.45<br>0.55 | | |
| PK012S | ½C | | | 0.06<br>0.09 | | | 0.17<br>0.23 | |
| PK021S | E | | | | | 0.15<br>0.25 | 0.35<br>0.45 | |
| PK022S | EL | | | | | 0.06<br>0.10 | 0.35<br>0.45 | |
| PK023S | ½E | | | | | 0.08<br>0.12 | 0.17<br>0.23 | |
| PK031S | F3 | 0.15<br>0.18 | 0.08<br>0.12 | | | | 0.10<br>0.13 | |
| PK041S | Cu-Te | | | | 0.030<br>0.045 | | | 0.035<br>0.045 |
| PK042S | ½Cu-Te | | | | 0.014<br>0.020 | | | 0.014<br>0.020 |
| PK043S | ¼Cu-Te | | | | 0.006<br>0.009 | | | 0.006<br>0.009 |
| PK049S | PbTeCu | | | | 0.03<br>0.045 | | | 0.03<br>0.045 |
| PK071S /<br>PK079S | Pb-Te | | | | | | | 0.035<br>0.045 |

Note: Lead alloys and their constituents suitable for submarine cables. The table shows the designation according to EN 50307, and the conventional names for the nearest related alloy.

## 12.3 Non-metric Conductor Size: kcmil

The unit is based on the concept of a circular Mil. A Mil is 1/1000's of an inch (= 0.0254 mm), and a circular Mil is the area of a circle with a diameter of 1 Mil. Now, 1 cmil = 0.0005064 $mm^2$ and 1 kcmil is 1000 cmil = 0.5064 $mm^2$. Sometimes the acronym MCM is used for kcmil.

| kcmils | $mm^2$ | kcmils | $Mm^2$ |
|---|---|---|---|
| 300 | 152.01 | 800 | 405.37 |
| 350 | 177.35 | 900 | 456.04 |
| 400 | 202.68 | 1000 | 506.71 |
| 450 | 228.02 | 1250 | 633.38 |
| 500 | 253.35 | 1500 | 760.06 |
| 600 | 304.03 | 1750 | 886.74 |
| 700 | 354.70 | 2000 | 1013.42 |
| 750 | 380.03 | | |

## 12.4 Non-metric Wire Diameter

In Anglo-Saxon markets the wire size is sometimes specified in non-metric units. Various "wire gauge" scales are deducted from the wire making process where the wire is drawn through consecutive dies each reducing the wire diameter. A higher wire gauge number indicates more wire drawing steps and a smaller wire. Some gauge scales are described here but there are more scales around.

AWG (American Wire Gauge)
Used in the United States since at least the 1880s for wires in all metals except iron and steel. Number 0000 wire is 0.4600 inch in diameter. The diameter of each succeeding size is 0.890525 times the diameter of the previous size.

BWG (Birmingham Wire Gage)
The steps are irregular. Departmental sanction by the United States government ended in 1914 but the scale is used widely even so.

SWG (Imperial Wire Gage, or British Standard Gage)
Legalized Standard Wire Gauge, Imperial Standard Wire Gauge, or in other countries, simply British Standard. Fixed by order of council August 23, 1883. It was constructed by improving the Birmingham wire gage. Made legal standard March 1, 1884.

| Gauge | AWG American Wire Gauge | BWG Birmingham Iron Wire | SWG Imperial Wire Gauge | AWG American Wire Gauge | BWG Birmingham Iron Wire | SWG Imperial Wire Gauge |
|---|---|---|---|---|---|---|
| | inch | inch | inch | mm | mm | mm |
| 7/0 | – | – | 0.5000 | – | – | 12.700 |
| 6/0 | 0.5800 | – | 0.4640 | 14.732 | – | 11.786 |
| 5/0 | 0.5165 | 0.500 | 0.4320 | 13.119 | 12.700 | 10.973 |
| 4/0 | 0.4600 | 0.454 | 0.4000 | 11.684 | 11.532 | 10.160 |
| 3/0 | 0.4096 | 0.425 | 0.3720 | 10.404 | 10.795 | 9.449 |
| 2/0 | 0.3648 | 0.380 | 0.3480 | 9.266 | 9.652 | 8.839 |
| 0 | 0.3249 | 0.340 | 0.3240 | 8.252 | 8.636 | 8.230 |
| 1 | 0.2893 | 0.300 | 0.3000 | 7.348 | 7.620 | 7.620 |
| 2 | 0.2576 | 0.284 | 0.2760 | 6.543 | 7.214 | 7.010 |
| 3 | 0.2294 | 0.259 | 0.2520 | 5.827 | 6.579 | 6.401 |
| 4 | 0.2043 | 0.238 | 0.2320 | 5.189 | 6.045 | 5.893 |
| 5 | 0.1819 | 0.220 | 0.2120 | 4.620 | 5.588 | 5.385 |
| 6 | 0.1620 | 0.203 | 0.1920 | 4.115 | 5.156 | 4.877 |
| 7 | 0.1443 | 0.180 | 0.1760 | 3.665 | 4.572 | 4.470 |
| 8 | 0.1285 | 0.165 | 0.1600 | 3.264 | 4.191 | 4.064 |
| 9 | 0.1144 | 0.148 | 0.1440 | 2.906 | 3.759 | 3.658 |
| 10 | 0.1019 | 0.134 | 0.1280 | 2.588 | 3.404 | 3.251 |
| 11 | 0.0907 | 0.120 | 0.1160 | 2.304 | 3.048 | 2.946 |
| 12 | 0.0808 | 0.109 | 0.1040 | 2.052 | 2.769 | 2.642 |
| 13 | 0.0720 | 0.095 | 0.0920 | 1.829 | 2.413 | 2.337 |
| 14 | 0.0641 | 0.083 | 0.0800 | 1.628 | 2.108 | 2.032 |
| 15 | 0.0571 | 0.072 | 0.0720 | 1.450 | 1.829 | 1.829 |
| 16 | 0.0508 | 0.065 | 0.0640 | 1.290 | 1.651 | 1.626 |
| 17 | 0.0453 | 0.058 | 0.0560 | 1.151 | 1.473 | 1.422 |
| 18 | 0.0403 | 0.049 | 0.0480 | 1.024 | 1.245 | 1.219 |
| 19 | 0.0359 | 0.042 | 0.0400 | 0.912 | 1.067 | 1.016 |
| 20 | 0.0320 | 0.035 | 0.0360 | 0.813 | 0.889 | 0.914 |
| 21 | 0.0285 | 0.032 | 0.0320 | 0.724 | 0.813 | 0.813 |

Note: Wire sizes under 0.7 mm diameter are omitted.

## 12.5 The Galvanic Series of Metals and Alloys in Seawater

| | |
|---|---|
| Magnesium and magnesium alloys | –1.60 to –1.63 |
| Zinc | –0.98 to –1.03 |
| Aluminum alloys | –0.76 to –1.00 |
| Mild steel | –0.60 to –0.71 |
| Wrought iron | –0.60 to –0.71 |
| Cast iron | –0.60 to –0.71 |
| Type 410 (13% chromium) stainless steel – active | –0.46 to –0.58 |
| Type 304 (18–8) stainless steel – active | –0.46 to –0.58 |
| Type 316 (18–8.3% Mo) stainless steel – active | –0.43 to –0.54 |
| Inconel (78% Ni; 13.5% Cr; 6% Fe) – active | –0.35 to –0.46 |
| Aluminum bronze (92%Cu; 8% Al) | –0.31 to –0.42 |
| Naval brass (60%Cu; 39%Zinc) | –0.30 to –0.40 |
| Yellow brass (65%Cu; 35%Zn) | –0.30 to –0.40 |
| Red brass (85%Cu; 15%Zn) | –0.30 to –0.40 |
| Tin | –0.31 to –0.33 |
| Copper | –0.30 to –0.57 |
| Lead-tin solder (50%–50%) | –0.28 to –0.37 |
| Admiralty brass (71%Cu; 28%Zn; 1%Sn) | –0.28 to –0.36 |
| Aluminum brass (76%Cu; 22%Zn; 2%Al) | –0.28 to –0.36 |
| Manganese bronze (58.5%Cu; 39%Zn; 1%Sn; 1%Fe; 0.3%Mn) | –0.27 to –0.34 |
| Silicon bronze (96%Cu; 0.80%Fe; 1.50%Zn; 2%Si; 0.75%Mn; 1.60%Sn) | –0.26 to –0.29 |
| Type 410 (13% chromium) stainless steel – passive | –0.26 to –0.35 |
| Lead | –0.19 to –0.25 |
| Inconel (78% Ni; 13.5% Cr; 6% Fe) – passive | –0.14 to –0.17 |
| Nickel 200 | –0.10 to –0.20 |
| Type 304 (18–8) stainless steel – passive | –0.05 to –0.10 |
| Monel 400 (70%Ni; 30%Cu) | –0.04 to –0.14 |
| Type 316 (18–8, 3% Mo) stainless steel – passive | 0.00 to –0.10 |
| Titanium | –0.05 to +0.06 |
| Platinium | +0.19 to +0.25 |

Note: Alloys are listed in order of the potential they exhibit in flowing seawater. Some alloys may become active and exhibit a potential near –0.5 V in low-velocity or poorly aerated water and at shielded areas.

## 12.6 Classification of Submarine Soil in Different Countries

| US Dept. of Agric. | Germany DIN 4022 | England BST 1377:1961 | Sweden (Atterberg) | Denmark |
|---|---|---|---|---|
| Cobbles (>75) | Stein (>60) | Stone (>60) | Block (>200) | |
| Coarse gravel (8–75) | Grobkies (20–60) | Coarse gravel (20–60) | Sten (20–200) | Sten (>20) |
| Fine gravel (2–8) | Mittelkies (6–20) | Medium gravel (6–20) | Grovgrus (6–20) | Grus (2–20) |
| Very coarse sand (1–2) | Feinkies (2–6) | Fine gravel (2–6) | Fingrus (2–6) | |
| Coarse sand (0.5–1) | Grobsand (0.6–2) | Coarse sand (0.6–2) | Grovsand (0.6–2) | Grovsand (0.2–2) |
| Medium sand (0.25–0.5) | Mittelsand (0.2–0.6) | Medium sand (0.2–0.6) | Mellansand (0.2–0.6) | |
| Fine sand (0.1–0.25) | Feinsand (0.06–0.2) | Fine sand (0.06–0.2) | Grovmo (0.06–0.2) | |
| Very fine sand (0.05–0.1) | Grobschluff (0.02–0.06) | Coarse silt (0.02–0.6) | Finmo (0.02–0.06) | Finsand (0.02–0.2) |
| | Mittelschluff (0.006–0.02) | Medium silt (0.006–0.02) | Grov mjäla (0.006–0.02) | |
| Silt (0.002–0.05) | Feinschluff (0.002–0.006) | Fine silt (0.002–0.006) | Fin mjäla (0.002–0.006) | Silt (0.002–0.02) |
| Clay (<0.002) | Ton (<0.002) | Clay (<0.002) | Ler (<0.002) | Ler (<0.002) |

Note: Denomination in local language. The grain size is given in mm.

Wentworth Scale

| Grain size | | Phi units | Sediment types |
|---|---|---|---|
| 4–64 | mm | –6 to –2 | Pebble |
| 2–4 | mm | –2 to –1 | Granule |
| 1–2 | mm | –1 to –0 | Very coarse sand |
| 0.5–1 | mm | 0–1 | Coarse sand |
| 250–500 | μm | 1–2 | Medium sand |
| 125–250 | μm | 2–3 | Fine sand |
| 63–125 | μm | 3–4 | Very fine sand |
| <63 | μm | >4 | Silt |

## 12.7 Non-metric Units

| | |
|---|---|
| 1 inch | = 25.4 mm |
| 1 foot (U.S. and British) | = 12 inches = 0.3048 m |
| 1 fathom | = 6 ft = 1.8288 m |
| 1 cable | = 219.4560 m |
| 1 nautical mile | = 1852 m |
| 1 lbs | = 0.45359 kg |
| 1 short ton | = 2000 lbs = 0.907185 MT |
| 1 long ton | = 1.016047 MT |
| 1 MT | = 1000 kg |
| 1 cubic inch | = 16.387 $cm^3$ |
| 1 cubic foot | = 0.028317 $m^3$ |
| 1 register ton | = 100 cubic foot = 2.8317 $m^3$ |
| 1 hectopascal | = 1 mb |
| 1 mm of mercury | = 1.3332 mb |
| 1 pound per square inch (psi) | = 0.06895 bar |
| 1 kn | = 1.852 km/h = 0.51444 m/s |
| 1 m/s | = 3.6 km/h = 1.94384 kn |

## 12.8 Tidal Terms

| English | English abbreviation | German | German abbreviation |
|---|---|---|---|
| Tides | | Gezeiten | |
| Height of tide | | Gezeitenhub | |
| Tidal streams | | Gezeitenstrom | |
| Highest astronomical tide | HAT | Höchstmöglicher Gezeiten-wasserstand | |
| High water | HW | Hochwasser | HW |
| High water heights | HW Hts. | Hochwasserhöhe | HWH |
| High water time | HW Time | Hochwasserzeit | HWZ |
| | | Höhe der Gezeit | H |
| Chart datum | CD | Kartennull, Kartendatum (Seekarten) | KN |
| Mean tide level | ML | Mittelwasser, Mittlerer Wasserstand | MW |
| Mean high water | MHW | Mittleres Hochwasser | MHW |
| Mean low water | MLW | Mittleres Niedrigwasser | MNW |
| Mean high water neaps | MHWN | Mittleres Nip-phochwasser | MNpHW |
| Mean low water neaps | MLWN | Mittleres Nipp-niedrigwasser | MNpNW |
| Mean high water springs | MHWS | Mittleres Springhochwasser | MSpHW |
| Mean low water springs | MLWS | Mittleres Spring-niedrigwasser | MSpNW |
| Lowest astronomical tide | LAT | Niedrigstmöglicher Gezeiten-wasserstand | |
| Low water | LW | Niedrigwasser | NW |
| Low water heights | LW Hts. | Niedrigwasserhöhe | NWH |
| Low water time | LW Time | Niedrigwasserzeit | NWZ |
| High water neaps | HWN | Nipphochwasser | NpHW |
| Low water neaps | NWN | Nippniedrigwasser | NpNW |
| Neap tides | Np | Nipptide | Np |
| Ordnance datum | OD | Normalnull | NN |
| Spring tides | Sp | Springgezeiten | Sp |
| High water springs | HWS | Springhochwasser | SpHW |
| Low water springs | LWS | Springniedrigwasser | SpNW |
| Slack water | | Stauwasser | |
| Admiralty tide tables | ATT | | |

# Reference

1. Deschamps L et al. (1980). Development in France of High Voltage Cables with Synthetic Insulation, Paper Cigré 21–06.

# Index



T. Worzyk, *Submarine Power Cables*, Power Systems,
DOI 10.1007/978-3-642-01270-9_BM2, 

## C



## D

**Batch number: 10755284**

Printed by Printforce, the Netherlands